MW01026940

Perfumery

Perfumery

Practice and Principles

Robert R. Calkin
Perfumery Training Consultant

J. Stephan Jellinek
Dragoco, Holzminden, Germany

A WILEY-INTERSCIENCE PUBLICATION
John Wiley & Sons, Inc.
NEW YORK / CHICHESTER / BRISBANE / TORONTO / SINGAPORE

This text is printed on acid-free paper.

To order books or for customer service please, call 1(800)-CALL-WILEY (225-5945).

Library of Congress Cataloging in Publication Data:

Calkin, Robert R.
Perfumery : practice and principles / Robert R. Calkin, J. Stephan Jellinek.
p. cm.
ISBN 0-471-58934-9 (alk. paper)
1. Perfumes. I. Jellinek, Joseph Stephan. II. Title.
TP983.C33 1994
668'.54—dc20 93-41844

Contents

PART IV ASPECTS OF CREATIVE PERFUMERY

PART V SCIENTIFIC FUNDAMENTALS

Appendixes

Preface

Some thirty years ago the perfumery profession was shaken by the commercialization of the gas chromatograph. In lectures, roundtable discussions, and private conversation hot debates centered around the question whether this analytical tool, by greatly simplifying the separation of complex mixtures of volatile materials, would make the perfumer redundant.

The initial stir soon calmed down, to be replaced by a feeling that the gas chromatograph, while highly useful to the analytical chemist and to quality control, would have little or no effect upon the perfumer's essential job, the creation of perfumes.

Today it is becoming increasingly clear that this business-as-usual second reaction was as misguided as the panicky initial response had been. For gradually but surely, the gas chromatograph—which soon expanded its scope of effectiveness by the successive introduction of capillary columns, the mass spectrometer, and quantitative head-space analysis techniques—has profoundly changed the perfumer's daily work.

In reviewing the changes, we concentrate upon the most salient ones: the erosion of secrecy and the intensification of competition, the acceleration of trends and of the trickle-down phenomenon, the rise of safety and environmental concerns, and the refinement of performance measurement.

Just one generation ago perfumers' work and the operations of the fragrance industry were steeped in secrecy. Access to the book of formulations was limited to a few trusted individuals. Perfumers kept to themselves any insights about the composition of famous perfumes which they might have acquired by dint of extensive matching efforts. The notion of giving customers information about the formulation of perfumes they were buying was anathema.

The fact that anyone armed with good GC/MS equipment and experienced in using this equipment can today, within days, find out a great deal about the formulation of any perfume has radically changed this climate. Formulas are still confidential, but the value of this confidentiality has been greatly diminished by the knowledge that, whenever it is to their advantage, customers and competitors can analyze most perfumes more or less precisely. The practice of giving formulas to major customers, either partially open with keys, or in toto in sealed envelopes to be opened under certain specified conditions, has long since ceased to be shocking.

This erosion of secrecy has logically led to a climate of more intensive competition, with results that are changing the face of the fragrance industry. Where formerly the industry offered good profit opportunities even to those companies that operated at less than peak efficiency, and the old artisan style of operation could survive alongside with the upcoming high technology approach, today only those fragrance houses are profitable that are either organized for peak efficiency or have specialized in optimally serving a specific niche of the market. In the brief span of five years (1986 to 1991) the share of world industry turnover accounted for by the ten largest fragrance and flavor suppliers increased from less than 50% to about two-thirds, and the trend continues. It would be an oversimplification to say that this development is due entirely to the gas chromatograph, but GC/MS certainly has played a major part in it.

It goes without saying that the growing pressures for efficiency in the industry are having marked effects also upon perfumers in their work, limiting their freedom of choosing raw materials (for the sake of keeping inventories in check), requiring them to accept judgments by evaluation boards or consumer panels on the quality and market appeal of their creations, and forcing them to make optimal use of the very tool that has played a major part in bringing about their new working environment, GC/MS. The task of exploring the composition of successful and trend-setting products in the market, which formerly occupied a major portion of the time of many perfumers, has become the work of specialized perfumer-chemist teams who conduct it more efficiently, leaving the perfumers free—but also forcing them—to concentrate their efforts upon the truly creative aspects of the job.

Gas chromatographic analysis today throws an increasingly clear light upon questions regarding the purity of natural perfumery materials. As a result perfurmers can now, if they are willing to pay the price, work with reliably pure materials; if they choose to use a commercial grade they know, more precisely than in the past, the material's degree of purity.

The history of perfumery has always been marked by evolutionary changes. Perfumers studied the perfumes they admired and built upon them, replacing their structural components by related yet different newer materials, shifting their center of gravity, introducing new nuances. At the heart of this process there was the quest for mastering the model perfume, and the modifications were like commentaries upon a classical text, highlighting certain of their features and showing their continued relevance in a changing world of fashions and styles. Mirroring the arduousness of the task of matching the model, the rhythm of this process was slow. When Madame Rochas, a commentary on Arpège, was launched in 1960, it took its place within an evolution of floral aldehydic perfumes that had continued without a break since 1921, the year of birth of Chanel No. 5. Fidji (1966) was a direct descendant of L'Air du Temps (1948).

Today the appearance of descendants of important fragrances is a nearly instantaneous process. The development and launching of mutations of and commentaries upon such perfumes occur at high intensity within very few years, even months after their launching, affecting not only the world of fine fragrances but also, thanks to the ubiquitous trickle-down phenomenon, deodorants and other toiletries, even laundry care and household products. As quickly as the trend has taken off, so quickly may it die down when the attention of the marketing community shifts to other models.

It would be wrong to attribute this change of pace to the GC/MS technique of matching; its origin lies, rather, in a transformation of marketing objectives. But in making near-instant matches possible, the GC/MS technique has provided the technical conditions that have made the dramatic speedup of derivation and of trickle-down possible.

The role that the gas chromatograph and mass spectrometry, along with a host of other high-powered new analytical techniques, have played in intensifying public concerns about the effects of fragrances and fragrance materials upon human health is perhaps not immediately obvious yet fundamental.

One generation ago public opinion about the wholesomeness of foods was dominated by ideas about main ingredients: Fresh fruits and vegetables are good for you, candy is not; proteins are good, too much fat is not. The same kind of thinking prevailed in the public mind about what is good or bad for the skin: Harsh solvents are bad, creams and oils are good. The knowledge that most foods and skin products contain, in addition to their main ingredients, small amounts of minor components including additives and trace residues of materials used in crop protection, processing, packaging, and so on, was confined to a small group of specialists.

Then, due to the rapid advances in analytical techniques, these specialists began finding more and more potentially harmful impurities even in foods that had been totally beyond suspicion: in fresh meat and beer, in baby food and milk including mothers' milk, even in the water we drink and the very air we breathe. It did not take long for these findings to catch the public's imagination and lawmakers' watchful attention. As a result we live in a climate of opinion today where public awareness of additives and impurities and their possible harmful effects has been raised to a level where it often overshadows thinking about the main effects of foods and beverages.

This climate has also colored the way the public perceives the products with which the perfumer deals: cosmetics and toiletries, detergents, and household products. Here too, additives and impurities are now seen as serious risk factors. One of the additives that has become salient and a cause for concern is fragrance. As a result perfumers today must constantly take into account the manifold and ever-changing safety-related requirements of the countries and of the specific customers for whom they create their fragrances.

In parallel to the awareness of possible health effects, concerns about environmental impact of materials and impurities have been greatly on the rise. The roots of this development do not lie in the advances in analytical chemistry but its course has been profoundly marked by these advances. It too is beginning to have a strong impact upon perfumers' work.

Turning now to the effects of analysis upon the objective quantitative measurement of fragrance performance, we broach a subject whose outlines are only just beginning to take shape and whose impact will be felt largely in the future.

Perfumery is the art of creating pleasurable and meaningful odor experiences. The nature of the experiences elicited by a fragrance in use is the measure of that fragrance's performance. The chain of events leading from the incorporation of a fragrance into a product to the experiencing of that fragrance by the product's user is complex and includes diverse chemical, physical, and neurophysical phenomena. In the past the only way perfumers could get information about what went on after they incorporated a perfumery material or a perfume in a product base was by smelling. Smelling is very imprecise when it comes to measuring intensity, a major aspect of performance. Moreover, being subjective, it is subject to all kinds of distortions. Since smelling registers only the mental experience that comes at the very end of the chain of events, it gives no clues about what went wrong at what point of the chain if performance is unsatisfactory.

Today quantitative head-space chromatography provides a means for monitoring objectively what is happening at an intermediate point in the chain. Coupled with improved techniques of extraction and liquid chromatography to identify chemical changes that perfume materials undergo in the product base, it can make perfumers far more efficient in arriving at highly performing fragrances than they had been in the past.

The idea for writing this book was the result of the authors' recognition of the profound changes in the nature of perfumery that are being brought about by recent advances in analytical techniques, and their awareness that these changes also call for a new approach to teaching perfumery.

One of us (R. R. C.) is daily observing, in his work as a teacher of perfumers, how the detailed knowledge of the structure of the great perfumes that today, for the first time, is becoming common knowledge, revolutionizes the teaching and learning of perfumery skills. The other (J. S. J.), who early in his career explored the physical aspects of fragrance performance and later centered his interest upon measuring and understanding human responses to fragrance, has seen his fields of interest similarly affected by the advances in analytical technique.

Writing this book in the midst of a period of rapid change, we are acutely aware that it cannot hope to offer the last word on any of the many topics it covers. We nevertheless hope that as an extensive review of an art in progress, it will prove useful not only as a textbook for prospective perfumers but also as an aid for those workers in the fragrance industry, the fragrance material industry, and the various industries of perfumed consumer products who want to improve their understanding of modern creative perfumery.

J. STEPHAN JELLINEK
ROBERT R. CALKIN

Holzminden, Germany
May 1994

Acknowledgments

The writing of this book has been supported in numerous ways by our colleagues at Dragoco and by the company's management. To them we owe our sincere gratitude. Without wanting to depreciate the contribution of the others, we should like to give special thanks to Dr. Hans-Ulrich Warnecke for his critical reading of Chapter 17, and to the perfumers who by their work on the great perfumes have provided a great deal of the substance of Chapter 12. Our thanks are due to Frank Rittler for much of the GC information used in the preparation of this chapter. Most of the perfumes mentioned in the text have also been worked on by the students of the Dragoco Perfumery School, and we are grateful to them for their enthusiasm in carrying out this work and for providing us with many ideas: Emily Coelho, Rolf Czyppull, Marc vom Ende, Jan Fockenbrock, Enrique Gomez, Yuko Ikeda, Veronika Kato, Styx Kwan, Susanne Multhoff, and Thomas Obrocki.

Outside Dragoco, we gratefully acknowledge the helpful exchange of thoughts with Dr. P. Müller and Dr. N. Neuner (Givaudan-Roure Forschung AG) on odor values, and their permission to use a chart from their work in Chapter 13, and the help of Birgit Kehmeier in calculating the vapor pressures for Tables 13.1–13.3 of that chapter.

R. R. C.
J. S. J.

Part I

Basic Skills and Techniques

1

What It Takes to Be a Perfumer

The decision to embark upon a career in perfumery is usually taken between the ages of 23 and 27. It is a serious decision because the job requires several years of demanding and rigorous training, a training so specialized as to be nearly useless for anything other than a career in the chosen field.

A basic understanding of creative perfumery is certainly helpful in a great number of other occupations such as sales, purchasing, production, application technology, quality control, research, and general management—but only within the rather narrow confines of the fragrance industry or of small departments within fragrance-using companies such as cosmetics and household products firms. And the perfumery skills that are useful in these functions are the ones that are learned within the first half-year of a good perfumery instruction course. Those who continue beyond this point should be absolutely sure of wanting to make creative perfumery their profession and reasonably sure of their chances of being successful at it. What qualifications can give them this certainty?

PHYSIOLOGICAL PREREQUISITES

Obviously a keen *sense of smell* is a primary prerequisite for any career in perfumery, yet, contrary to popular belief, exceptional keenness is not required. Skill in perfumery is related more to what the brain does with odor perceptions than to the perceptions themselves.

A given person does not necessarily have a uniformly keen or poor sense of smell. It often happens that someone is highly sensitive to certain odor notes, while perceiving others relatively poorly or not at all. This phenomenon is known to physiologists as *partial anosmia.* Perfumers should not have too pronounced partial anosmias, although they do not have to be entirely free from the condition. Some very skillful perfumers are unable to smell certain perfumery materials (usually musks or woody odorants of high molecular weight) in pure form, yet perfectly capable of noting the effects of these materials—a strange phenomenon that is familiar to psychophysicists who have observed that the olfactory threshold for a substance in mixtures is often different than for the same substance alone.

Olfactory sensibility is subject to change over time, perhaps related in part to hormonal influences. Many perfumers have moreover found that their sensitivity varies with the time of day, so they adjust their working habits accordingly. Surprisingly this question has not, to our knowledge, been systematically investigated; major changes in the course of the day have, however, been well established in the related field of dosage-response ratios for drugs.

There is some truth to the popular belief that perfumers should not be smokers. The evidence that at least heavy smokers have reduced sensitivity for at least certain types of odors is now irrefutable (Gilbert and Wysocki 1987). Nevertheless, a number of the great perfumers of the past have been heavy smokers. It is often claimed that perfumers, in order to keep their sense of smell in top condition, must avoid strongly seasoned foods. The experience of individual perfumers appears to vary rather widely in this respect; every perfumer should find out for him- or herself what foods or beverages, if any, adversely affect performance.

It has been clearly demonstrated that olfactory sensitivity declines with age and that this decline becomes particularly pronounced after age 60 (Gilbert and Wysocki 1987). However, not all perfumers retire at 60. Many have performed successfully throughout their 60s and even in their 70s. Whatever losses in sensitivity they suffered, they compensated by their long experience. Once more, then: High sensitivity is necessary, but exceptional sensitivity is not.

Discrimination ability, that is, the capability of recognizing small differences in odor quality, can to some extent be learned. However, there are innate, permanent differences between people in the extent to which they can acquire this ability, and the perfumer must be very good at it. The beginning student will come to know his or her own potential ability only after some time.

The same is true for odor *memory,* for it can also be acquired or at least trained. There are large individual differences here, it is certainly more a mental faculty than an attribute of the olfactory organ. An excellent odor memory is indispensable to the perfumer. The student perfumer who after six to nine months of training still has difficulties in recognizing the 200 or so main perfumery materials should seriously question whether it is wise to continue training.

PREREQUISITES OF PERSONALITY

Successful perfumers represent a rather wide range of personality types, but is seems fair to say that a balanced disposition is optimal: (1) a predilection for the patient, persistent, and essentially lonely effort that the creation of a fragrance requires, balanced by the gregariousness needed for the give-and-take that is at the heart of a successful team effort; (2) the reliance upon the perfumer's own aesthetic intuitions in their pursuit of the creative effort, tempered by a sensitivity toward the expectations of clients and consumers who may have very different tastes; and (3) the high standards and perfectionism that it takes to create fragrances of exceptional quality, coupled with the realism needed to live within the limitations inherent in nearly every practical assignment.

An eagerness for recognition and success is important as a motivating force, but it should not be too central a motivation lest the perfumer get distracted by a quest for facile successes or discouraged by frequent disappointments. Disappointments are inevitable because of the way major projects are handled: Nearly always, several perfume suppliers are involved and, within each supplier firm, several perfumers, but in the end only one can win. The closer one comes to winning, the keener the disappointment if the perfume is not selected. The fact that the decision is usually based on undefinable, subjective considerations makes it harder yet to accept.

The person who has the ambition to achieve great things for humankind should not become a perfumer. Successes in perfumery may yield considerable financial rewards and may lead to recognition within a small circle of professionals. But from the perspective of human history they are insignificant.

The person to whom power over people is an important motivator should seriously question whether perfumery is the right profession. Certainly a perfumer's decision about using or not using a certain raw material may have major consequences to the supplier of that material.

Certainly a successful perfumer may become a power to reckon with within his or her firm. But the moment such power ceases to be a side effect of success and becomes a major motivator, the perfumer's professional performance suffers.

The good perfumer's strongest motivation is enthusiasm, born out of the sheer joy that he or she derives from doing the job. The "appetite that is brought on by the foretaste of discovery" must, in Stravinsky's (1942) words, be in the perfumer "as habitual and periodic, if not as constant, as a natural need." The perfumer must know the "pleasure of creation . . . that is . . . inseparable from . . . the very act of putting [his or her] work on paper" even if this act may entail a great deal of effort, and even struggle. The joy of creation is the source of the perfumer's persistence in the face of difficulties, the medicine against the frustrations that are bound to come up time and again and the secret of the perfumer's enthusiasm. It is also an essential component of creativity.

PREREQUISITES OF CREATIVITY

We have no ambition to enrich the vast existing literature on creativity by yet another contribution. We prefer to restrict ourselves to a few personal observations with regard to this subject in perfumery.

Creativity in perfumery, as in any art, has a great deal to do with the ability to make observations that are spontaneous, direct, and unfettered by any traditional views on how things should be; an ability to discover new potentials in a material that has already been used a thousand times; to note the flash of the unexpected that comes from a chance combination, in a certain proportion, of familiar odors; to recognize a special character in a perfume material that runs counter to what its chemical structure would have led one to expect. It is perhaps not by chance, then, that many good perfumers are also good photographers. Both arts involve an aptitude for fresh observation, for breaking through the crust of traditional viewing patterns.

In apparent contrast to what we have just said, creativity also presupposes a great deal of experience and knowledge in the chosen field. To turn the spark of an idea into an aesthetically sound perfume takes a great deal of skill; even conceiving the idea requires a solid base of knowledge. Spontaneity and enthusiasm alone are not enough. We do not know of any case in which a perfumer with less than a few years of solid experience has created a truly good perfume.

Inspiration in the sense of a guiding vision prior to the work is not necessary. The creation of a perfume may well start, to borrow another phrase from Stravinsky (1942, p. 55), as a kind of "grubbing about" and the spark, the jolt, or even a succession of sparks comes only as the work proceeds.

Apart from the creativity that expresses itself in novel odor accords and effects, there is also the inventiveness that leads to new solutions to such problems as masking the base note of a particularly unpleasant-smelling functional product or achieving outstanding diffusion and lasting power in a low-price soap perfume. The question whether the two—we might call them *aesthetic* and *technical* creativity—are different in kind has long divided the perfumery profession. Suffice it to say that technical creativity is, from a commercial point of view, every bit as valuable as the aesthetic kind.

Aesthetic creativity is not a guarantee for success. Commercial success has to do with sensing how much newness and unconventionality any given assignment will tolerate. We will discuss this point in more depth in the chapter on the Perfumer and Marketing.

In summary, we can say that in the perfumer's work, creativity is manifested largely in the power of imagination that enables one to place the idea of an odor that does not exist yet in the mind's eye, in the spontaneity of perception that enables one to discover ever new aspects and potentials in long-familiar perfume materials, in the cleverness to conceive of new approaches to old problems, and in the dedication and skill that it takes to turn a vision or a discovery into a good perfume.

EDUCATIONAL PREREQUISITES

A key question is: Should the perfumer have studied chemistry? Some perfumers' reply is a firm and unequivocal "no." This reaction is found primarily among those who concern themselves largely or exclusively with "alcoholic perfumery" (i.e., the creation of compositions for perfumes, colognes, etc.). A prominent exponent of this wing, Edmond Roudnitska, said: "Perfumers are chemists no more than is the painter who manipulates chemical colors. In itself, composing a perfume has nothing to do with chemistry" (Moreno et al. 1974, p. 71). "The Compositeur must not let himself be influenced by systematic thoughts. *Only by considering each odor by itself and in its rapport with the other odors, without any preconceived idea, will he make the best use of it*" (Moreno et al. 1974, p. 109).

We consider Roudniska's view to be largely true. The perfumer who thinks too "systematically" might well be blind to or at least underrate the vast differences among, for instance, methyl, amyl, hexyl, and benzyl salicylate or between vetivert oil and vetiveryl acetate. In using rose oxide, the perfumer might limit him or herself too much to the rose complex, and using citral, to the citrus complex. The total openness with which perfumers should approach the odor of perfume materials can be constrained by thought patterns that are too much chemically oriented. Nevertheless, we feel that an understanding of chemical processes can greatly facilitate the perfumer's job, especially when it comes to perfuming chemically active functional products, and we therefore strongly recommend that perfumers, except perhaps those specializing exclusively in alcoholic perfumery, should have a background in organic chemistry and be familiar with the fundamentals of physical chemistry.

A knowledge of botany enriches the understanding of natural perfume materials and their production but is not essential to the perfumer's job. An interest in the physiology of olfaction and the mechanisms of odor detection is intellectually enriching, but at the current state of knowledge this discipline contributes little to the practice of perfumery.

The study of odor psychology and of what has of late been called "Aroma-chology,"* of human responses to odors, is a different matter. There is much here that is related to perfumers' work. The same may be said about the anthropology of odor, the study of the differing roles and meanings of odors in different cultures. Unfortunately, publications in these fields are dispersed over a great number of specialized journals that are not readily accessible to perfumers.

In a later chapter of this book, we will attempt to give a glimpse of the information available in the area of the psychophysics of odor. Psychophysics is the science that seeks to measure the perception of sensory inputs in a quantitative manner and to arrive at general laws of perception.

As to language skills: We regard fluency in written English as indispensable, for this is the language of the major international trade literature. Perfumers working for companies that are active in the international market should be fluent also in spoken English, since this will enable them to enter into direct exchanges with most foreign customers. To the perfumer with strong interests in fine perfumery, we recommend proficiency at least in written French, for in this sector French perfumers and publications in French have always played a key role.

*A service mark of the Olfactory Research Fund.

TESTING FOR APTITUDE

At Dragoco, where a formal perfumery training course has been conducted for some time, candidates for the course are screened in a test that includes an odor recognition test and a series of triangle tests. In the odor recognition tests, candidates are asked to identify different odors presented on a smelling blotter. The odor selections range from fruits, spices, and other food-related odors to the odors of leather, cigar boxes, and paint thinner. The score is indicative of a basic ability to perceive and to remember odors as well as odor awareness and articulateness. Women nearly always outperform men in this test. An exceptionally high score is not a prerequisite for the perfumery training course, though a very low score may well indicate serious deficiencies that may affect one's career potential.

The triangle test is a measure of odor discrimination. The candidate is presented, in random order, with three blotters, of which two are identical and the third is slightly different. The task is to indicate the odd blotter. This test can be designed to range from easy to very difficult. It is statistically most powerful if the difference between the paired blotters is such that the odd one is correctly identified by about 50% of the candidates.

Questions may also be included in which the candidate is asked to pick the best and the poorest out of a group of, say, lavender or ylang oils or rose bases. Sometimes totally inexperienced candidates score surprisingly well in this test, which measures innate taste and sense of quality.

In these tests it is not only the numerical score that counts. The candidate's enthusiasm and involvement in tackling the tasks, the joyful excitement he or she may show when a test odor calls forth an old memory or when the sample is of a particularly fine material: These are crucial, if nonquantifiable, indicators of aptitude.

2

The Student Perfumer Today

Like any other creative art the art of perfumery depends upon experience and technique as well as upon inspiration. Experience and technique can only be acquired by an immense amount of patient study and hard work, frequently beset by disappointment and frustration. Yet for the talented and enthusiastic student the obstacles are more than offset by the sense of discovery and excitement that surrounds the work. The training will usually involve either working as an apprentice (often as a compounder) to a senior perfumer or formal study in one of the perfumery schools. The apprentice perfumer will probably be required to spend some time gaining firsthand experience in other departments of the company, in related areas such as production, the application of perfumes in different functional products, product evaluation and marketing, analytical techniques, and quality control. There is little room today for the "ivory tower" approach to perfumery; a perfumer is seen as part of a team of experts working toward the success of a company.

It was frequently the case in the past that a perfumer could enter the profession with little or no formal training, often having worked first as either a chemist or a laboratory assistant in a related area of the industry. Perfumery knowledge and experience came through a long and often difficult period of self-training, with only as much information and help as could occasionally be obtained from senior colleagues. This approach was rightly based on the principle that there is no substitute for the hard work that a student must do in order to acquire the necessary knowledge of raw materials and basic perfumery formulation to enable him or her to work successfully on new creations, and to establish a personal perfumery style. However, the role of the perfumer today, in an industry that has become increasingly technical and market orientated, demands a breadth of knowledge and experience that can only be achieved within a reasonable period of time by intense and well-planned instruction.

In the past, details of perfumery formulation were the jealously guarded secrets of perfumers who had spent a lifetime acquiring them. A student perfumer would be expected to spend many months working alone, with little help, trying to recreate one of the great perfumes of the past, while doing hundreds of simple experiments with a handful of traditional raw materials so as to establish the best relationships between them. Such work is of course still of value in the training and subsequent work of the perfumer, but modern techniques have made it possible to shorten this method of training so that the student can progress far more quickly than in the past. (A similar revolution has taken place in the field of music, where modern teaching methods have resulted in a generation of young musicians whose technical and interpretive ability is quite astounding.)

One of the factors that has triggered the change in the teaching of perfumery has been the introduction of gas chromatography. Any company today, for an initial investment of little more than the annual salary of one perfumer, can obtain a vast amount of information about the formulation of existing perfumes. It would be foolish to deny the impact that this availability of information can have on modern methods of perfumery training.

It is here that the experience of the teacher can be of the greatest importance. Too much teaching and the giving of too much information can be as harmful as too little, for young perfumers must make their own discoveries and develop their own personalities. Merely to present students with a bookful of formulas does not make them perfumers. Indeed it is more likely to inhibit their creativity and spirit of inquiry. Although the teacher may set the course that students should follow, and give judiciously of his or her experience and formulations, the work of becoming perfumers must be done by the students themselves.

In perfumery, as in painting, photography, or music, there are no set rules of technique. At best we may formulate a number of general principles that will guide students in their work and to which they will continually return. No two perfumers work in precisely the same way, but discipline and technique there must be if students are to make the progress that will give them a sense of fulfillment and the determination to carry on. In our own experience the most important role of the teacher or senior perfumer in the training of young perfumers is to maintain their enthusiasm, instilling in them a spirit of inquiry, while providing them with the discipline and sufficient information to allow them to make the rapid progress that will fuel their excitement and creative self-confidence.

From the very beginning of their studies and throughout their careers, perfumers must practice the discipline of olfactory training. Each morning they must set aside time for testing their knowledge of materials against a number of samples, selected by their teacher, a colleague, or assistant. Even senior perfumers can find this difficult when faced with a test taken from the entire inventory. As with any other skill, this expertise requires both practice to achieve and practice to maintain. The learning of materials will be discussed in more detail in a later chapter.

Good laboratory training is another aspect of young perfumers' education. The proper labeling of samples, writing of formulas, and maintenance of the laboratory and laboratory equipment must be established right from the beginning before bad habits become ingrained. Perfumers today must retain an enormous amount of information and must be able to work fast and under pressure. The more orderly the recording of information, the freer the perfumer to think creatively. Young perfumers must be encouraged to have a good intellectual grasp of their work, and this should be reflected in the way in which they write out their formula. As we will show in a later chapter, a perfume is not just a random mixture of materials that somehow come together with a beautiful effect; a perfume has a well-defined structure. This may of course vary from one perfume type to another, and the past eighty or more years has seen the evolution of a number of different styles. Students must always be aware of the structural framework within which they are working.

Matching provides one of the best ways of learning perfumery. It is as important for perfumery students as it is for young musicians to study the works of the great composers. But too much matching can have a paralyzing effect on the minds of aspiring perfumers. They risk becoming creatively lazy and dependent, and neither their imaginations nor initiatives are allowed to grow. Today, however, much of the drudgery has been taken out of matching by the use of gas chromatography, allowing young perfumers to become familiar with the composition of many of the great perfumes within a comparatively short period of time. But advanced students must learn to work with GC information intelligently in order to gain a genuine insight into the way in which perfumes have been constructed. They should never concern themselves only with slavish imitation. The technique of matching will be discussed in greater detail in a later chapter.

Most perfumers today would agree that the foundation of perfumers' training should be in fine perfumery and in the great perfumes, past and present. All art forms are partly derivative in their style and technique. Each individual artist is influenced by what has gone before, adding something new of his or her own creative ideas and personality. Similarly in perfumery one can trace the development of modern families of perfumes back to the perfumes of fifty or sixty years ago, some of which are as popular today as when they were first created and will probably outlast many of their lesser progeny. It is therefore an essential part of the training and repertoire of young perfumers to work within all the main families of perfumes, to understand their underlying structure, and to explore the ways in which these can be modified to produce new fragrances, often in types of application and at a cost far removed from the original. Within this tradition there remains plenty of scope for students to express their own imagination.

Many of the perfumes for functional products are derived from successful fine fragrances. The adaptation of these fragrances for different types of products and the creation of original fragrances for similar applications is a technique that must be learned, following on from the study of fine perfumery.

In summary, although there are no hard and fast rules in the construction of a perfume, there are certain principles that need to be adhered to. We will repeatedly refer to these principles in the pages that follow.

1. A profound knowledge of the raw materials is the basis of perfumery technique and inspiration.
2. A perfume is not just a random collection of pleasantly smelling raw materials. It is the result of a precise system of structures within the formulation.
3. A perfume's structure is based upon:
   a. The precise olfactory relationship between individual ingredients, known as the "perfumery accord."
   b. The relationship between simplicity and complexity.
   c. The balance between materials of different volatility suitable for the product for which the perfume is intended.
4. A perfume needs to fulfill certain technical requirements, such as chemical stability in use.

In addition students must always remember that perfumery is an evolving tradition. Although they may ultimately go on to produce great original creations, much of their early work must be spent mastering the techniques handed down by previous generations of perfumers. Many of these will serve students well in the creation of their own perfumes.

A word of warning for young perfumers: One of the most difficult things for students is to sustain enthusiasm and patience over what is inevitably a long and difficult period of study. The work is both introspective and competitive. All too often students may be faced with disappointment after months of work. Self-confidence can easily be destroyed, particularly when a student is allowed to charge ahead without building up the solid foundation in olfactory memory and perfumery technique on which success is based. If a perfumer ceases to enjoy his or her work, and every new project carries with it the threat of failure rather than the hope of success, the most likely cause lies in the insecurity of the technique.

Above all students must learn to enjoy not only their work but the life of a perfumer. Making time for other creative interests outside of work is equally important, helping one to relax one's mind between the periods of intense concentration that are needed to achieve success.

3

The Technique of Smelling

In perfumery as in all specialized fields good basic techniques are essential. The perfumer should acquire a correct approach to smelling right from the start. In our remarks on this topic, we will quote from the views of Paul Jellinek (1954) and Edmond Roudnitska (1962, 1991). Both wrote as experienced practicians, though with very different backgrounds. Jellinek had worked in the essential oil industry and dealt with the entire spectrum of perfumery, including the perfuming of functional products of all kinds. Roudnitska, after initially holding a position in the soap and detergent industry, established a studio in the hills near Grasse and worked as a free-lance perfumer nearly exclusively in fine perfumery. Despite their different backgrounds the agreement between the two authors is remarkable, as is the fact that their remarks, written several decades ago, are still fully valid today.

THE WORKPLACE

On the workplace Jellinek noted: "It is self-evident that smelling in a well-ventilated room is much easier than in a smelly laboratory or in a place full of soap dust. The perfumer, therefore, should have at his disposal a separate room, which may house his desk, his books, his formulary, etc., but from which samples of aromatics or of strongly smelling soap, as well as all laboratory work should be strictly kept out." Roudnitska had this to say: "An olfactory test should be undertaken in odorless, tempered air of natural humidity and in quiet surroundings. Perfect concentration definitely requires solitude and quiet. The substance to be tested is also hard to smell in excessively cold or dry air, or in a draught."

Yet, despite the last injunction, air conditioning is necessary in hot climates. Where air conditioning is not needed, adequate air circulation must be provided in the perfumer's working area by a direct supply of outside air, by creating a slight overpressure in the working area such that the air flows out from it into the surrounding areas of the building. Short of having Roudnitska's rare chance of working at a good distance from any manufacturing operations, a certain degree of odor impregnation of the air cannot be avoided. The casual visitor, highly aware of this background odor, is surprised that it does not interfere with the perfumer's demanding job. Due to the strange phenomenon of long-term adaptation, the perfumer does not notice the steady background odor that has become a habitual part of the working atmosphere; only when the perfumer returns to work from a lengthy absence does he or she suddenly become conscious of it.

Where outside air quality is a problem, charcoal or other suitable filters must be built into the ventilation system. Some companies have tried to create ideal conditions for critical smelling by constructing smelling cabins with metal or glass walls, supplied with air that is deodorized and kept at constant, optimal temperature and humidity and with a provision for rapid air exchange between tests. In our experience such cabins were hardly ever used by perfumers. The sterility of the cabin creates a kind of stress that makes work difficult. The room where the perfumer does his or her smelling should be functional but furnished and decorated so as to make the occupant feel at ease. The odor-free cabins usually end up being used for the evaluation of space odorants or detergent and fabric softener perfumes on the fabric after drying. For such testing, they are ideal.

THE SAMPLES

Both Roudnitska and Jellinek advise doing one's smelling with dilute solutions in order to avoid odor overload and fatigue. Sound as this advise may be in theory, it raises the question what solvents to use for dilution. For work in alcoholic perfumery, alcohol is the obvious one. Care must be taken to let it evaporate entirely before bringing the blotter to the nose, since inhaling alcohol temporarily deadens the sense of smell. For projects related to other product fields, the choice of diluent becomes problematic because each diluent affects the odor performance of a perfume compound in its own way. Ideally a diluent should be chosen that is very similar or identical to the product base to be perfumed, but in practice this can become extremely laborious. Nearly all perfumers do smell raw materials and compounds in their neat, undiluted form, avoiding overload and fatigue by not dipping the blotters too deeply and confining their actual smelling to the brief span of a few inhalations. This should be conducted with the intense concentration of a karate fighter. Casual, thoughtless smelling must be avoided at all times.

One should never sniff directly at a bottle, for this inevitably deadens the sense of smell for some time. Neither should one sniff at a crystalline material in its undissolved form, for the odor of such materials is often greatly distorted by traces of impurities adsorbed at the surface. Sniffing at finely powdered materials such as vanillin involves the added risk of inhaling small crystals directly into the nose, and thus destroying one's ability to smell for a considerable period to come.

SMELLING BLOTTERS

Roudnitska recommends blotters 18 cm in length; Jellinek suggests that they be at least 10 cm. Today blotters are commonly 13 to 15 cm long, and ½ to 1 cm wide. Roudnitska suggests using 1-cm wide blotters folded lengthwise so that they form a kind of groove and bend less readily; he also recommends that the blotters be tapered, which facilitates dipping them into small bottle openings and also minimizes the amount of material to be examined. The paper must be unsized. Roudnitska recommends a grade of 180 g/sq cm; Jellinek observes that for odor analysis, the thinner the blotter the better, since "it allows the different smelling phases to be recognized, as it holds the more volatile aromatics less tenaciously." For presentation of a finished perfume to a customer, on the other hand, he recommends heavier, more absorbent paper because "it holds the full composition better than a thinner one."

SMELLING

The nose adapts quickly to unchanging odor. Active smelling is therefore always a race against time, an attempt to collect the maximum amount of information and of impressions possible in the brief span of time before perception fades. It must always be done with the most intense concentration. The surroundings should be quiet. The body should be relaxed and comfortable. Closing the eyes helps to avoid visual distraction. All attention is focused upon the sensation of the odor. If identification is the objective, Jellinek suggests posing oneself precise questions, to be answered by "yes" or "no," in the manner of a Twenty Questions game. If, for example, one has recognized a floral note and wants to identify it, one asks oneself, before bringing the blotter to the nose: "Is it jasmin?" If the answer is no, one continues: "Is it rose? Lily of the valley? Honeysuckle? Hyacinth?" and so forth, until an identification is made. If there is doubt in answering some of these questions, it may be because one does not remember with sufficient clarity the odor of the flower in question. It then helps to go back to the shelf and sniff a few good bases of the desired type (not directly from the bottle!) to refresh one's memory.

Both Roudnitska and Jellinek stress the desirability of keeping written notes of one's odor impressions. As Jellinek states: "By writing them down [one] is forced to concentration and to express [one's] sensations in a clear manner. Lucidity of expression demands a clear recognition of the impression received." Writing down observations not only forces one to concentrate and to smell with awareness, it also aids in recalling the odor later. Although the memory of an odor that one really knows stays with one for a lifetime, the mental image of a new odor can fade rapidly, hence, the importance, emphasized also by Roudnitska, of writing down one's immediate impressions.

To avoid smelling fatigue, the sessions must include rest intervals during which the mucosa can recover. Jellinek suggests using the break for studying, Roudnitska recommends: "Wherever possible, the examiner should go into the fresh air and take a brief turn out of doors after each series of inhalations and before beginning the next one." In an industrial setting this practice may meet with some raised eyebrows. We have found running up a flight of stairs a good alternative. By stimulating deep breathing and activating the blood circulation, it seems to help clear the nose.

4

Perfumery Raw Materials

The raw materials used in perfumery are traditionally divided, according to their origin, into naturals and synthetics. Although in reality this division is not nearly as clear-cut as it appears at first, it is a convenient starting point for this brief review.

NATURALS

We designate as natural all materials that are obtained from natural sources by the application of physical separation techniques such as distillation and extraction. Natural products have been used for many thousands of years as the raw materials of perfumery. Entire plants, flowers, fruits, seeds, leaves, as well as woods, roots, and the resins they exude, are all sources of fragrance materials. Similarly the scent glands of animals such as the civet cat and the musk deer have been used since early civilization to provide perfume for humans.

Originally in the ancient civilizations of Egypt and Greece fragrant plant material was macerated in the unguent oils used for softening and perfuming the skin or in wine. In the Middle Ages Arab scientists developed the technique of **steam distillation** by which concentrated **essential oils** could be prepared by passing steam through moistened plant materials and then cooling and collecting the distillate. These oils, originally intended and used for medicinal purposes, in time also came to be used in perfumery.

A clear distinction between medicine and perfumery was not drawn until around 1800. In the previous centuries the main purpose of pleasing and beneficial fragrances was seen as warding off and curing the illnesses caused by noxious emanations. Today steam distillation is still one of the most important methods of producing natural perfumery materials. Some essential oils, such as those which occur in the skins of citrus fruits, are more usually produced by direct **expression** from the plant material.

In the eighteenth century, concentrated alcohol, prepared by the distillation of fermentation alcohol, became available. This made possible the preparation of **tinctures** prepared by the maceration of plant or animal materials in such alcohol. These played a major role in nineteenth-century perfumery. Today, for practical reasons and because of the cost involved, their use is entirely restricted to fine perfumery. They are still used in some of the perfumes that were created in the early decades of our century.

A breakthrough in the extraction of floral perfumes occurred in the late eighteenth century with the introduction of the technique known as **enfleurage**. In this method the flowers were placed either directly on or in close proximity to a layer of animal fat which absorbed the fragrance. The fats were then washed with alcohol which was subsequently removed by distillation to produce a concentrated product known as an *absolute,* often referred to *an absolute from pomade or chassis*. Owing to the very high labor costs involved, the enfleurage method of extraction has almost entirely disappeared, being used only rarely for the extraction of tuberose.

By the early part of the twentieth century, pure grades of volatile hydrocarbon solvents such as benzene and hexane became available through progress in petroleum-refining methods. They were found to be very useful for the **extraction** of fragrant plants and plant materials. If the plant material extracted is rich in waxes (as is generally the case with flowers, stems, and leaves), these are also taken up in the extract. After careful removal of the volatile solvent by distillation, a waxy **concrete** remains behind. This is then washed with alcohol to separate the fragrance materials, which are soluble in alcohol, from the insoluble waxes. An **absolute** is then produced by the removal of the alcohol by distillation, usually under reduced pressure. Certain plant materials that contain no water, such as resins or dried leaves and mosses, may be extracted directly with alcohol. The extracts obtained—often sticky, viscous, and resiny—are called **resinoids**.

Recently, owing partly to restrictions in the levels of solvent residues allowed in finished extracts, particularly for the flavor industry, alternative solvent systems have come into use for extraction. Interesting results are being obtained using liquid carbon dioxide under high pressure in place of conventional solvents. Materials produced in this way are quite distinct in character and provide a new challenge for the perfumer.

Many other processes can be applied to concretes, absolutes, resinoids, and essential oils to obtain products with special characteristics. For example, **decolorized** products may be obtained by extraction with appropriate solvents to eliminate highly colored components. Colorless products may be obtained by **fractional molecular distillation** under vacuum. Yet another type of product is obtained by **vacuum codistillation** using a suitable solvent. Many essential oils, particularly citrus oils, contain high levels of insoluble terpene hydrocarbons, which can be removed by fractional distillation or by countercurrent extraction to produce concentrated or **terpeneless** oils.

With fractional distillation, single aroma chemicals can also be obtained, in more or less highly purified form, from essential oils. These are designated as **isolates**. In the days when synthetic organic chemistry was less versatile and powerful than it is today, this was the only way to obtain many aroma chemicals. Even today it is in some cases, such as citronellal or cedrol, the most economical way.

SYNTHETICS

Today there are several thousand synthetically produced aroma chemicals potentially available to the perfumer. Many of these—for example, vanillin, the rose oxides, and the damascones—were first discovered in nature and subsequently synthesized. Others are the fruit of a chemist's imagination and have never been found in nature. Of course not all are of equal value to the perfumer, and the number that are widely used runs into the hundreds rather than into the thousands.

One of the earliest perfumery materials to be produced industrially was benzaldehyde, prepared from toluene in 1866. In 1868 coumarin was first synthesized, soon to be followed by heliotropin, ionone, and vanillin. The first nitro musks appeared in 1888, and amyl salicylate in 1898. Since then, for the last hundred years, the history of perfumery has been dominated by the creation of new aromatic chemicals.

For most new aroma chemicals the starting point of synthesis was the hydrocarbons obtained from petroleum refining or the monoterpenes obtained from turpentine. The synthesis of the aldehydes provided the inspiration for such perfumes as Chanel No. 5 and Arpège, and phenylethyl alcohol became the indispensable rosy material throughout perfumery. Other milestones include the determination of the structure of farnesol (the basis for the sesquiterpenes) by the Nobel Prize winner Leopold Ruzicka, working for Chuit & Naef in Geneva, in the early 1920s, followed by his synthesis in 1928 of cyclopentadecanolide (Exaltolide), the first of the macrocyclic musks. Closer to our own time, methyl dihydrojasmonate (Hedione) was introduced in the early 1960s, followed by such materials as Galaxolide, Vertenex, Brahmanol, and Iso E super.

THE GRAY AREA BETWEEN NATURALS AND SYNTHETICS

The dividing line between natural and synthetic materials is far from clear-cut. The very basis of the distinction is questionable, since it assumes that no chemical changes occur in the course of processes such as steam distillation and extraction. It has long been known that this assumption is not correct. In addition there are gray areas between the two categories that can best be demonstrated by a few examples.

Geraniol, a substance of defined chemical structure, may be extracted from a natural source such as palmarosa oil or produced synthetically from pinene. Rigorous purification of geraniol from either source will produce the chemically pure substance. Depending upon its origin, this should then be designated as either natural (from palmarosa) or synthetic (from pinene), although the two grades can be distinguished only by very sophisticated analysis.

The "perfumery grade" qualities of geraniol from the two sources actually have considerable differences in odor value because neither is chemically pure. Geraniol from a natural source contains small amounts of many other materials carried over from the source material. Synthetically produced geraniol is a reaction mixture, frequently containing a large proportion of nerol and other substances not found in the natural product.

Either of these "perfumery grade" geraniols can be acetylated to geranyl acetate. The acetates, having been prepared by a chemical reaction, should be designated as synthetic in either case, yet the differences between the two grades are as distinct as they are with the two geraniols.

Many natural products such as labdanum, cedarwood, and clary sage are fractionated and chemically treated (in this or in the reverse order) to produce derivatives of exceedingly complex composition which, for the perfumer's purposes, behave as naturals. These can be refined down to more-or-less chemically pure single materials whose status as natural or synthetic depends upon whether they are present, in the same form, in the starting material.

Natural materials are frequently "cut" or "touched up" with synthetic materials either to standardize the quality and the analytical constants between different crops (the explanation usually given) or simply to reduce the cost. Such products, although frequently and regrettably sold as pure, are in extreme cases little more than compounds.

One of the salutary effects of the widespread adoption of quality control techniques based on gas chromatography has been to inject clarity into this formerly somewhat murky area. Thanks to these techniques, the purchaser is now in a position to know to what extent a so-called natural material is pure. As a result genuinely and reliably pure essential oils, absolutes and resinoids are again available in the market today. Synthetic essential oils that are honestly designated as such may be thought of as quasi-natural materials rather than as bases.

It is part of the perfumer's job to be aware of the differences between such materials and to learn how different qualities of a given material perform in perfumery compositions. Young perfumers should begin by knowing pure natural products and high-quality synthetics, and by working with them before going on to work with the so-called commercial qualities.

5

The Learning and Classification of Raw Materials

Building a foundation in perfumery technique begins with a thorough knowledge of the raw materials. Raw materials are to the perfumer what colors are to the painter, or what words are to the poet. One can attain mastery of perfumery only insofar as one knows the necessary materials; they are both the tools of the perfumer's trade and the perfumer's inspiration.

The novice perfumer may well feel daunted by the hundreds of bottles containing strange and often unpleasant smelling materials that line the laboratory shelves. But for the talented student the task of learning to identify them is in fact less difficult than it may seem at first. However, this is only the first step along the way to having a real knowledge of the materials, a knowledge which can only come by actively working with them over a period of time. The professional perfumer is time and again presented with new materials coming either from plant products or from the result of chemical research. On each occasion the perfumer must become acquainted with their odor qualities and their performance in use.

As in learning a foreign language, the first stage in learning the materials—the vocabulary of perfumery—necessitates repeated smelling and testing. This task can be made simpler, and in the long term more effective, if it is approached in a systematic way, namely one based on an understanding of how human memory works.

Getting to know the raw materials is not a training of the nose or the olfactory receptors (it is doubtful that the receptors can be trained at all), it is the training of the thought processes that provide the link between the perception of an odor and one's ability to recognize it and give it a name. These processes, as all mental processes, are based on a complex network of associations. Each new odor learned, like each new word in our vocabulary, is integrated into an existing framework of meanings and olfactory associations. Fortunately, the more a student knows, the easier it becomes to add new materials to the existing memory bank.

The process of recognition of a smell begins from the moment it comes in contact with the receptor cells of the nasal cavity. Although the precise manner in which these cells respond is not fully understood, it is generally accepted that the sensory input from an individual chemical is related to the different aspects of its molecular structure. For instance, all materials with a phenylethyl radical, within certain limits of molecular size, are related in odor, as are most nitriles and salicylates. The impulses generated by the structure of the molecule are processed by the brain to produce the cognitive experience that we call smell. In the case of complicated mixtures such as natural products and perfumes, the amount of incoming information that needs to be processed is considerable. It is believed that the initial integration of the olfactory stimulus and its cognitive recognition is carried out by the right hemisphere of the brain, whereas its name recognition is dependent on the further activity of the left hemisphere. This may account for the frustrating "tip-of-my-tongue" syndrome—I know this, but I cannot give it a name. Paul Jellinek's admonition, never to smell a substance without full concentration, is particularly apt here, since getting to know an odor is mentally demanding. With practice, the connections between perception, recognition, and naming will be speeded up for the student to the point where a known material may be identified almost instantaneously. Even so, it is interesting to note that even to the fully trained perfumer the ability to give names to visible objects seems to require much less of a mental effort.

Complicated mixtures may sometimes take on an identity which is no more difficult to recognize than a single chemical. Perfumes such as Anais Anais or Giorgio are as instantly recognizable as phenylethyl alcohol or anisaldehyde. Similarly lavender oil or geranium have a unique and memorable identity. Indeed, in our experience, students often can more easily identify these complicated mixtures than single chemical materials. Just why some perfumes have such strong identity is not fully understood, but, as will be discussed later, much of the skill of the perfumer lies in the ability to achieve such an identity in a composition.

As we noted earlier, it is important for the novice student to learn the raw materials of perfumery in a systematic way. The student should first be introduced to no more than about 50 materials selected by importance but including as wide a range of odor types as possible. (A suggested list is given later in this chapter). The student should be encouraged to write down some descriptive remarks about these odors, using whatever associations come to mind. These odors may conjure up memories from the past, such as the smell of grandmother's cupboards, hot potato pancakes, or freshly dug-up roots of old trees. Comparisons between different materials should be made and similarities noted. At this stage such descriptions reflect the individual associations of the student, and neither the teacher nor fellow students can judge them as being objectively right or wrong. In this way the student consciously builds up a network of associations to assist in remembering the odors. Once the 50 materials have been learned the number can gradually be increased over a period of several weeks to include all the 162 given in the list.

At this early stage students should only be introduced to materials of the highest quality. Both natural and synthetic materials exist in a number of different qualities depending on their source of origin, method of production, and purity. It is important for students to acquire a clear olfactory impression of the finest materials before going on to the use of the more "commercial" qualities.

During the first few months of training students should be encouraged to start making a classification of odor types. One such system of classification is given here, but it is important for students to go through the mental process of working out their own classifications, even if the teacher has to make some later adjustments. Although learning by association and classification is a useful way for getting to know the materials and to think about them, students must never be restricted by too rigid a framework of classification. They must be encouraged to open their minds to all the different facets of a material and to be constantly looking for new associations.

In learning to recognize materials, students should also conduct evaporation tests to see how long each material retains its characteristic note on the smelling strip. Some materials last no more than a few hours, while others are still recognizable after several days. Such results should be carefully recorded, and the records will form the basis for a student's later understanding of the importance of comparative volatilities in the structure of a perfume.

Students should also be made aware of differences in olfactory strength between various groups of materials. Many materials, such as the aliphatic aldehydes and some of the animalic products, will be strong even when diluted at 10% or 1%. Others, such as benzyl salicylate, are difficult to smell at all until they become more familiar. But these differences can sometimes be deceptive, and materials that may not smell exceptionally powerful on their own can have a significant effect in a composition, even when used at very low concentrations. The performance of individual odorants, rather than their apparent strength, is best learned by experiencing their behavior in different types of formulation. Sometimes, for example, we may use vanillin at 10% in a formula, and yet 0.01% in another may have a remarkable effect. Young perfumers will often benefit from not "knowing" that a material cannot be used at more than 1%. Being less inhibited about its use than the more experienced perfumer, they may come up with something excitingly original.

A further insight into the learning process, and one that is closely related to perfumery, comes from our experience of learning about wines. Although by drinking them regularly we may learn much about their individual character, by far the most effective method of becoming a genuine connoisseur comes from attending wine tastings at which a number of vintages are contrasted and compared, one against the other. This technique of learning by association, similarity, and contrast is one that works particularly well in perfumery.

Every morning students should be given unmarked smelling strips of ten or more materials selected from those to which they have already been introduced. Those that a student fails to recognize should be discussed with the teacher and fellow students and notes made in the classification. At times the materials selected by the teacher should be as different in odor type as possible; at other times they should be taken from the same olfactory group, such as the rose notes, woody materials, or chemically related materials such as the acetates. Once the basic materials have been learned and understood, it is easier to add new materials. In this way the intelligent student will even be able to name materials not previously encountered. For example, the student may identify phenylethyl phenylacetate from its characteristic rose note combined with the honey note typical of the phenylacetates.

We have now reached a point at which we can define the three stages of odor familiarity: First, we identify an odor relatively, then we know it absolutely, and finally we know it actively. In the first stage of familiarity we are able to identify an odor by our description of it, allowing our minds to search through our stored-up associations and classification. For example, given a sample of benzoin, our first impression may simply be "sweet." This in itself doesn't get us very far, since it encompasses far too large a group of odor impressions. From here we may go on to describe it as "chocolatelike with a touch of incense." Finally, we may distinguish it from tolu in which we would expect to find a slight leathery character, or from peru balsam which lacks the incense note (an observant student will probably begin by noting first the differences in color!). Most perfumers, as well as such people as professional tea tasters and wine experts, have developed highly sophisticated and differentiated vocabularies to describe these small differences in odor between similar products. One team of perfumers were able to communicate among themselves the idea of a particular type of green note by the phrase "smelling like elephant's feet." The language of perfumery is unfortunately so poor that sometimes we need to invent such metaphors in order to communicate.

We may be said to know a material absolutely when our recognition of it becomes immediate without our having to think of a description. We recognize the odor like an old and familiar friend. We recognize benzoin simply because it smells of benzoin. But even an old and familiar friend may exhibit personality or behavioral characteristics that surprise us. Of course now and then our memory may not perform too well and we need to allow our minds to go through the partly conscious and partly subconscious processes of classification and recognition. But once remembered we can use the material, and the others on the basic list, as a point of departure for describing and memorizing other materials. When first shown tolu we may think of it as similar to benzoin, which we know absolutely, but a little more leathery. Phenylethyl formate is seen as related to phenylethyl acetate but with the typical sharpness that we found in geranyl formate.

Finally, we learn to know a material actively. To continue with our analogy, being able to recognize someone instantly in the street is not the same as really knowing him or her. You know a person only if you have observed that person's behavior in different situations. The same is true of perfumery materials. One really comes to know a material only by actively working with it. For this reason students should be encouraged to work with materials within a few weeks of beginning their studies, beginning with simple accords before going on to the floral bases and the work of matching the great perfumes of the past and present. (This work will be discussed more fully in a later chapter.) As with people, the adventure of getting to know a material never comes to a close. Even the experienced perfumer who has worked with a material over decades may discover new characteristics by using it in new combinations and applications. The more the perfumer keeps an open mind for such discoveries, the more creative he or she will be. Having an active knowledge of the ways in which materials work in combination and in different types of formulas is what perfumery is all about.

So far we have considered memory from the point of view of recognition and identification. There is yet another side to memory that is concerned with recall and mental imagery. Here the perfumer is at a natural disadvantage when compared to the artist or musician. Most of us have more or less well developed senses of visual and auditory recall. The ability to see familiar objects in our mind's eye is one that we put to constant use, and the trained artist can reproduce on paper realistic images from memory. Similarly a musician can be trained to know precisely the sound that will be produced from the notes of a musical score, as might be heard in an actual performance. However, the ability to recall smell is usually much less developed. Can we really recall the smell of a rose in the same way that we can recall a color or a melody? We may be able to describe the smell and recapture all the associations that go with it, but can we experience a concrete olfactory image of it in our minds?

What we can do, without too much difficulty, is to compare something that we smell against our recollection of it, and determine whether or not it is the same. This ability is particularly important in assessing the quality of raw materials, yet in fact this is just another form of recognition (do we recognize the sample as being the same as before?). There is no true olfactory imagery where no external stimulus is involved.

There are good reasons why human beings have failed to develop an ability for olfactory recall compared with other forms of perception. Auditory recall is an essential part of our ability to use sound as a means of communication and language, and visual recall, including our ability to see things as they might be rather than as they are is basic to our purposeful manipulation of objects. But olfactory recall does not serve such everyday needs, it has no apparent value in the evolution of our species.

Still many perfumers do claim some degree of olfactory recall, and this ability should be developed in young perfumers and actively pursued in practice. A useful exercise is to deliberately keep alive the olfactory impression of a material for as long as possible after ceasing to smell it and then compare the retained image against the original. Of course this requires great concentration, but it will help to develop an intuitive sense of how materials work in combination. Some perfumers claim to have developed their talent for olfactory recall to a point where much of their creative work is done in their imagination.

THE BASIC RAW MATERIALS AND THEIR CLASSIFICATION

The following is a list of the 162 natural and synthetic materials that it is suggested students should learn to identify within the first six months of their studies. It is intended to cover a wide range of odor types. These materials are mainly used in fine perfumery. (Materials marked with an asterisk form the list of 50 materials to be learned first.)

NATURALS

Ambrette seed
*Armoise
Basil
Bay
*Benzoin Siam
*Bergamot
Birch tar
Camomile Roman
Cardamon
Cassis bourgeons (base)
*Castoreum
*Cedarwood Virginian
Celery seed
Cinnamon leaf
*Cinnamon bark
Cistus oil
Civet
Clary sage
Clove bud
Coriander
Costus (base)
Cumin
*Estragon
*Galbanum oil
*Geranium Bourbon
Guaiacwood
Iris concrete
Jasmin absolute
*Labdanum extract
*Lavender
Lavandin
*Lemon
Lemongrass
Lime West Indian
Mandarin
Mimosa absolute
Neroli
Nutmeg
*Oakmoss absolute
Olibanum extract
Opoponax extract
Orange sweet
*Patchouli
Pepper
Peppermint
Peru balsam oil
*Petitgrain Paraguay
Pimento
Rose oil
Rose absolute
*Rosemary
Rosewood
*Sandalwood East Indian
Styrax oil
Tagete
Tonka absolute
Tuberose absolute
Thyme
Vanilla absolute
*Vetyver Bourbon
Violet leaf absolute
*Ylang extra

SYNTHETICS

Acetophenone
Aldehyde C10
*Aldehyde C11 undecylenic
Aldehyde C12 lauric
*Aldehyde C12 MNA
*Aldehyde C14 (gamma-undecalactone)
Aldehyde C16
Aldehyde C18 (gamma-nonalactone)
Allyl cyclohexyl propionate
Ambroxan
Amyl cinnamic aldehyde
*Amyl salicylate
Anisaldehyde
Aurantiol
Benzaldehyde
*Benzyl acetate
Benzyl salicylate
Brahmanol
Calone
Cashmeran
Cedramber
Cedryl acetate
Cinnamic alcohol
Citral
Citronellal
*Citronellol
Citronellyl acetate
Coumarin
Cyclamen aldehyde
Cyclopentadecanolide
Damascone beta
*Dihydromyrcenol
Dimethyl benzyl carbinyl acetate
Diphenyl oxide
*Ethyl phenylacetate
Ethyl vanillin
*Eugenol
Evernyl
Frambinone
*Galaxolide
gamma-Decalactone
Geraniol
*Geranyl acetate
Geranyl formate
Geranyl nitrile
Greenyl acetate
Hedione
Helional
Heliotropin
**cis*-3-Hexenyl acetate
cis-3-Hexenyl salicylate
*Hexyl cinnamic aldehyde
*Hexyl salicylate
*Hivertal
*Hydroxycitronellal
*Indol
Ionone alpha
*Isobornyl acetate
*Iso butyl quinoline
Isoeugenol
Iso E super
Isogalbanate
cis-Jasmone
Lilial
*Linalool
*Linalyl acetate
Lyral
Maltol
*Methyl anthranilate
Methyl benzoate
Methyl cinnamate
Methyl chavicol
*Methyl ionone gamma
Methyl napthyl ketone
Methyl octine carbonate

Methyl salicylate
Musk ketone
Musk T
Paracresyl acetate
Phenoxyethyl isobutyrate
*Phenylacetaldehyde
Phenylacetic acid
Phenylacetaldehyde dimethyl acetal
Phenylethyl acetate
*Phenylethyl alcohol
Phenylethyl dimethyl carbinol
Phenylethyl phenylacetate
Phenylpropyl alcohol
Rosalva
Rosatol
*Rose oxide
Sandela
*Styrallyl acetate
*Terpineol
Tonalid
*Vanillin
Vertacetal
*Vertofix
Vetiveryl acetate
Vertenex (PTBCHA)

TABLE 5.1 Classification by Odor of 162 Materials

Class	Naturals	Synthetics and (Ingredients) of Naturals	Class	Class/ Specific Character
ALDEHYDIC		Aldehyde C10		Orange
Aldehydes		Aldehyde C11 undecylenic		
		Aldehyde C12 lauric		Soapy
		Aldehyde C12 MNA		(Pine)
		Citronellal		Cintronellal
(Nitrile)		Geranyl nitrile	*Citrus*	
(Alcohol)		Rosalva		Rose
AMBER				
	Cistus oil			
	Labdanum extract		*Resinous*	
		Cedramber	*Woody*	Cedarwood
		Iso E Super	*Woody*	Patchouli
		Ambroxan	*Woody*	Dry wood
ANIMALIC (*see also* Amber and Musk)				
	Civet			Fecal
	Castoreum		*Leather*	
		Indol	*Floral*	Jasmin
		Paracresol	*Floral*	Narcisse
		Phenylacetic acid	*Floral*	Honey
	Costus		*Iris*	Hair
	Cumin		*Spicy*	Sweaty
ANISIC				
	Basil	Methyl chavicol		
	Estragon	Methyl chavicol		

TABLE 5.1 (*Continued*)

Class	Naturals	Synthetics and (Ingredients) of Naturals	Class	Class/ Specific Character
AROMATIC-HERBAL (*see also* Anisic)				
	Armoise	(Thujone)	*Camphor*	
	Roman camomile			Hay
	Clary sage		*Linalool*	
	Thyme			
BALSAMIC				
	Benzoin Siam	(Vanillin)	*Resinous*	*Sweet*
	Peru balsam oil	(Vanillin)	*Sweet*	
	Styrax oil	(Phenylpropyl alcohol)	*Cinnamic*	
		Cinnamic alcohol	*Cinnamic*	*Floral*
		Phenylpropyl alcohol	*Cinnamic*	*Floral*
		Methyl cinnamate	*Cinnamic*	*Floral*
		Benzyl salicylate	*Salicylate*	*Floral*
CAMPHOR-CINEOL				
	Armoise	(Camphor, thujone)	*Herbal*	*Aromatic*
	Cardamon	(Camphor)	*Spicy*	*Seed*
	Lavandin	(Camphor, cineol)	*Lavender*	*Linalool*
	Rosemary	(Camphor, cineol)	*Lavender*	Eucalyptus
CINNAMIC				
	Cinammon bark	(Cinnamic aldehyde)	*Spice*	
	Cinnamon leaf	(Eugenol)	*Spice*	
		Cinnamic alcohol	*Floral*	*Balsamic*
		Phenylpropyl alcohol	*Floral*	*Balsamic*
		Methyl cinnamate	*Floral*	*Balsamic*
	Styrax oil		*Balsamic*	
		Amyl cinnamic aldehyde	*Floral*	Jasmin
		Hexyl cinnamic aldehyde	*Floral*	Jasmin

CITRUS				
		Citral		Lemon
	Lemon	(Citral)	Fruity	Fresh
	Lime distilled	(Terpineol)	Fruity	Fresh
	Sweet orange		Fruity	Fresh
	Mandarin		Fruity	Fresh
	Bergamot	(Linalool, L. acetate)	*Fresh*	*Linalool*
	Lemongrass	(Citral)		
		Citronellal	*Aldehydic*	Citronella
		Geranyl nitrile	*Aldehydic*	Lemon
FLORAL				
Almond blossom				
		Anisaldehyde		Lilac
		Benzaldehyde		Almond
Carnation		Eugenol	*Spicy*	Clove
		Isoeugenol		
Gardenia		Styrallyl acetate	*Fresh*	
Hyacinth		Phenylacetaldehyde	*Green*	Rose
		Phenylacetaldehyde dimethyl acetal	*Green*	Lilac
Honey				
		Ethyl phenylacetate		
		Phenylethyl phenylacetate		Rose
		Phenylacetic acid	*Animalic*	Rose
Jasmin	Jasmin absolute			
		Benzyl acetate		*Fresh*
		Hedione		
		Jasmal		Mushroom

TABLE 5.1 (*Continued*)

Class	Naturals	Synthetics and (Ingredients) of Naturals	Class	Class/ Specific Character
FLORAL (*Continued*)				
		Amyl cinnamic aldehyde	*Cinnamic*	
		Hexyl cinnamic aldehyde	*Cinnamic*	Muguet
		Indol	*Animalic*	
	Ylang extra		*Linalool*	
Lilac				
		Terpineol		Household
		Anisaldehyde	*Floral*	Almond
		Heliotropin	*Sweet*	
Mimosa				
	Mimosa absolute			
		Acetophenone		
Muguet				
		Hydroxycitronellal		
		Lilial		
		Lyral		
		Cyclamen aldehyde		
Narcisse-Jonquille		Methyl cinnamate	*Cinnamic*	
		Paracresol	*Animalic*	
Neroli-Orange Blossom		Methyl anthranilate		
		Aurantiol		Schiff base
		Methyl napthyl ketone		
	Neroli, Petitgrain		*Fresh*	
Rose	Rose absolute	Phenylethyl alcohol		
	Rose oil	Citronellol		
		Citonellyl acetate		Muguet
		Geraniol		
		Phenylethyl acetate	*Fresh*	Fruity

		P.E. dimethyl carbinol		Raspberry
		DMBCA		
		beta-Damascone	*Fruity*	
		Phenoxyethyl isobutyrate	*Fruity*	
		Rosatol		
		Phenylethyl phenylacetate		Honey
	Geranium		*Minty*	
		Geranyl acetate	*Fresh*	
		Geranyl formate		
		Rose oxide	*Green*	Metallic
		Diphenyl oxide		
Trefle (salicylate)				
		Amyl salicylate		
		Benzyl salicylate	*Balsamic*	
		Hexyl salicylate		Azalea
		cis-3-Hexenyl salicylate		Green
Tuberose				
	Tuberose absolute			
		Aurantiol		
		Methyl anthranilate		
		Methyl benzoate		Ylang
		Methyl salicylate		Wintergreen
		Aldehyde C18	*Fruity*	
		gamma-Decalactone	*Fruity*	
Violet				
	Violet leaf absolute		*Green*	
		Methyl octine carbonate	*Green*	
		Ionone alpha	*Iris*	
		Methyl ionone	*Iris*	*Woody*
	Iris concrete		*Iris*	
		Heliotopin	*Sweet*	Lilac
Ylang				
		Ylang extra		
		Methyl benzoate		

TABLE 5.1 (*Continued*)

Class	Naturals	Synthetics and (Ingredients) of Naturals	Class	Class/ Specific Character
FLORAL FRESH				
	Neroli		Orange blossom	
		Linalool	*Linalool*	*Woody*
		Benzyl acetate	Jasmin	
		Citronellyl acetate	Rose	Muguet
		Geranyl acetate	Rose	Geranium
		Phenylethyl acetate	Rose	Fruity
		Styrallyl acetate	Gardenia	Rhubarb
FLORAL BALSAMIC				
		Cinnamic alcohol		
		Phenylpropyl alcohol		
		Benzyl salicylate	Trefle	
FRESH (*see also* Floral fresh and Citrus)				
	Bergamot	(Linalyl acetate)	*Citrus*	*Linalool*
	Petitgrain	(Linalyl acetate)		Eau de cologne
		Linalyl acetate		
		Isobornyl acetate	*Pine*	
		Dihydromyrcenol	*Linalool*	
		Vertenex (PTBCHA)	*Woody*	
FRUITY				
	Citrus oils (see Citrus)			
		Aldehyde C14		Peach
		Aldehyde C16		Strawberry

		Aldehyde C18	*Tuberose*	Coconut
		gamma-Decalactone	*Tuberose*	Peach
		beta-Damascone	*Rose*	Apple
	Tagete			Apple
		Allyl cyclohexyl propionate		Pineapple
		Isogalbanate	*Green*	Pineapple
		Phenoxyethyl isobutyrate	*Rose*	Mirabelle
		Helional	*Watermelon*	
		Frambinone	*Sweet*	Raspberry
		Cashmeran	*Musk*	*Sweet*
	Cassis bourgeons			Cats
GREEN				
	Galbanum			Earthy
		Hivertal		Leafy
		cis-3-Hexenyl acetate		Cut grass
		Isogalbanate	*Fruity*	Pineapple
		Phenylacetaldehyde	*Hyacinth*	*Rose*
		PADMA	*Lilac*	*Hyacinth*
		Vertacetal	Sweaty	Rhubarb
	Violet leaf absolute		*Violet*	
		Methyl octine carbonate	*Violet*	
		Rose oxide	*Rose*	Metallic
HERBAL (*see* Aromatic, Anisic, Lavender)				
IRIS				
	Iris concrete		*Violet*	
		Ionone alpha	*Violet*	
		Methyl iononee	*Violet*	*Woody*
		Vertenex (PTBCHA)	*Woody*	*Fresh*
	Costus		*Animalic*	

TABLE 5.1 (*Continued*)

Class	Naturals	Synthetics and (Ingredients) of Naturals	Class	Class/ Specific Character
LAVENDER				
	Lavender 40/42	(Linalool, Linalyl acetate)		
	Lavandin	(Linalool, Linalyl acetate)	*Camphor*	
	Rosemary		*Camphor*	Eucalyptus
LEATHER				
	Castoreum		*Animalic*	
	Birch tar			Smoky
	Labdanum		*Amber*	
	Cistus		*Amber*	
		Isobutyl quinoline		Earthy
LINALOOL				
		Linalool	*Fresh*	*Woody*
		Dihydromyrcenol	*Fresh*	
(Essential oils containing linalool)				
	Bergamot	(Linalyl acetate)	*Citrus*	*Fresh*
	Clary sage		*Aromatic*	
	Coriander		*Spicy*	*Seed*
	Lavender	(Linalyl acetate)	*Lavender*	
	Lavandin	(Linalyl acetate)	*Camphor*	
	Rosewood	(Terpineol)	*Woody*	
	Ylang	(Benzyl acetate)	*Floral*	
MINTY	Peppermint			
	Geranium Bourbon		*Rose*	
MOSSY	Oakmoss			Algal
		Evernyl		Dusty
MUSK				
		Galaxolide		
		Musk T		

		Tonalid		
		Musk ketone		
		Cyclopentadecanolide		
	Ambrette seed	(Ambrettolide)	*Seed*	
		Cashmeran	*Sweet*	*Fruity*
PATCHOULI				
	Patchouli			Pepper
PINE				
		Isobornyl acetate	*Fresh*	
		Aldehyde C12 MNA	*Aldehydic*	
RESINS				
	Benzoin Siam		*Sweet*	*Balsamic*
	Olibanum extract			Insence
	Opoponax extract			
	Labdanum extract		*Amber*	*Leather*
	Peru balsam oil		*Sweet*	*Balsamic*
SEEDS				
	Ambrette		*Musk*	
	Cardamon		*Spice*	*Camphor*
	Celery			Terpenic
	Coriander		*Linalool*	*Spice*
	Cumin		*Spice*	*Animalic*
	Pepper		*Spice*	Terpenic
SPICY				
	Cinnamon bark	(Cinnamic aldehyde)		
	Cinnamon leaf			
		Eugenol	*Carnation*	
	Clove bud	(Eugenol)		
	Pimento berry	(Eugenol)		

TABLE 5.1 (*Continued*)

Class	Naturals	Synthetics and (Ingredients) of Naturals	Class	Class/ Specific Character
SPICY (*Continued*)				
	Bay			
	Nutmeg			Terpenic
	Pepper			Terpenic
	Cardamon		*Camphor*	*Seed*
	Coriander		*Linalool*	*Seed*
	Cumin		*Seed*	Sweaty
SWEET (Powdery)				
	Vanilla absolute	(Vanillin)		
		Vanillin		
		Ethyl vanillin		
	Benzoin Siam	(Vanillin)	*Balsamic*	
	Peru balsam oil	(Vanillin)	*Balsamic*	
		Coumarin		Hay
	Tonka	(Coumarin)		
		Heliotropin	*Lilac*	*Violet*
		Frambinone	*Fruity*	
		Cashmeran	*Musk*	*Fruity*
WATERMELON—CUCUMBER				
		Calone		Marine
		Helional		
	Violet leaf absolute		*Green*	
	Mimosa absolute		*Floral*	

WOODY				
Cedarwood				
	Cedarwood Virginian			Pencils
		Cedryl acetate		
		Vertofix		
		Cedramber	*Amber*	
Guaiacwood				
	Guaiacwood			Smoky
Rosewood				
	Rosewood	(Linalool)	*Linalool*	
Sandalwood				
	Sandalwood E.I.			
		Sandela		
		Brahmanol		Musk
Vetyver				
	Vetyver Bourbon			Earthy
	Vetiveryl acetate			
Woody-Amber		Iso E Super		
		Cedramber		Cedarwood
		Ambroxan		Dry wood
Woody-Iris		Methyl ionone	*Violet*	
		Vertenex (PTBCHA)	*Fresh*	

6

The Floral Accords

The floral accords form an essential part of the perfumer's repertoire. They are valuable not only as building blocks, or bases, in the creation of all types of perfume but also as fragrances in their own right. With some modification the floral accords can be used as single floral notes for functional products such as soaps, cosmetics, and household products. They also form an excellent starting point for student perfumers in the formulation of simple compounds.

The fragrance of flowers, and their association with our sense of aesthetic beauty, has always held a special place in our appreciation of smell. To faithfully recapture their scents by distillation, enfleurage, and solvent extraction has been one of the continuing objectives of the fragrance industry since the Middle Ages. The crude chemical analysis of the resultant products that took place during the first half of the century gave little information as to their composition that could not at least be guessed at by an experienced perfumer. A major breakthrough came as a result of the introduction of gas chromatography. This resulted in the identification of many hundreds of components, many of which were previously unknown. However, most extracted products, although some the most beautiful materials used in perfumery, fail to capture the full character of the natural flower because of the chemical changes that take place during the extraction process. More recently head space gas chromatography has made possible the direct analysis of the fragrance as it leaves the plant, adding considerably to our knowledge of the composition of the fragrance that we actually smell in nature. Many commercially available products now exist based on this information.

Another approach to the reproduction of floral notes has come from the work of the creative perfumer based on the use of synthetic materials whose olfactory character may be similar in some respect to that of the flower being copied, whether or not they are known to occur in nature. Many of the floral bases created in this way, although frequently but pale reflections of the flowers themselves, have come to be among the most widely used building blocks of perfumery.

The compounding notes that follow provide a guide to the formulation of some of the most important types of floral bases used in perfumery. The "basic formula" that is given in each case contains materials that represent the main olfactory elements necessary for the recreation of the fragrance. Establishing a balance between these is a useful introduction to understanding the way in which the various types of floral are composed.

It is important when beginning work on a floral base to start with as simple a formula as possible, containing only sufficient materials to establish the essential character of the note. This basic accord can later be embellished by the addition of modifying materials and naturals. The most effective floral bases, particularly when designed to be used as a major part of a perfume formula, are often comparatively simple mixtures containing no more than 15, and often fewer materials. More complicated bases, when used in this way or in combination with other florals, may often be the cause of "muddiness" and general lack of impact in the final product.

Having arrived at their own preferred proportions within the accord, in keeping with the target floral type, students may then experiment with other materials or alternatives from within the main categories, initially restricting themselves to a maximum of two or three from each. At this stage it is important to be as adventurous as possible in the choice of materials. Not every trial will be successful, but with imagination and courage interesting and original formulas will result. To make a rose with 50% of phenylethyl alcohol is no problem, but to begin with 20% of diphenyl oxide is something of a challenge.

Many floral notes overlap in the types of material from which they can be made. The same eight materials—phenylethyl alcohol, hydroxycitronellal, benzyl acetate, phenylactaldehyde, citronellol, hexyl cinnamic aldehyde, terpineol, and indol—mixed together in different proportions can produce either a jasmin, a lilac, a muguet, or a hyacinth. The list also includes three materials that occur in nearly all rose compounds. If used together in a formula, bases that contain only such generally used materials are liable to lose their identity giving merely a nondescript floral character with little or no personality. It is important therefore, when working on a floral note, to include as large a proportion of highly characteristic materials as possible, which will carry the note through into the final product. However pleasing and well balanced a base may smell on its own, if it fails to hold together and perform in a composition, it is of little value.

Many hyacinth compounds are particularly disappointing in this respect, with the fine floral character being reduced to no more than a "green" effect when used in combination with other floral notes in the finished perfume. The most successful accords, whether they be those that are found in floral bases or those on which the great perfumes are founded, seem to be virtually indestructible wherever they occur. With experience students will learn the ways in which the various types of floral base perform in different types of composition. Most perfumers choose to work with only a limited number of bases, whose performance they are thoroughly familiar with and understand.

An exception to the rule of simplicity may sometimes be made in the creation of a base designed to duplicate, at a lower price, the effect of the natural flower extract. Such bases, many of which benefit from the results of analysis, are frequently used as auxiliary products to bring a "natural" character to the composition, rather than as part of the basic structure.

It is not unusual in the composition of a perfume to use not only an individual floral material such as phenylethyl alcohol as part of the main structure of the formula but to back this up by a more complex rose base as well a little of the natural flower product. This principle of simplicity within the main structure of the perfume, and increasing complexity within the lesser components, is one to which we will return in later chapters.

ROSE—COMPOUNDING NOTES

Basic Formula	Selected Variants and Modifiers
Phenylethyl alcohol and esters	
Phenylethyl alcohol	
	Phenylethyl acetate
	Phenylethyl esters
	Phenoxyethyl isobutyrate
Rose alcohols and esters	
Citronellol	Rhodinol
	Geraniol
	Nerol
Geranyl acetate	Geranyl esters
	Citronellyl esters
	Tetrahydrogeraniol
	Rholiate
Other rose chemicals	
	Rose oxides
	Isodamascone
	Damascones
	Orthomethoxy benzyl ethyl ether
Floral modifiers	
Ionone alpha	Beta ionone
	Linalool
	Lyral
	Hedione
Green notes	
Phenylacetaldehyde	Phenylacetaldehyde dimethyl acetal
	Methyl heptine carbonate
	Violet leaf absolute
	cis-3-Hexanol and esters
	Hivertal
	Isocyclocitral
	Nonadienal
	Methyl cyclo citrone
	Vertral
Aldehydic notes	
Aldehyde C11 undecylenic	Aldehyde C8
	Aldehyde C9
	Rosalva
	Alcohol C10
	Muguet aldehyde

Basic Formula	Selected Variants and Modifiers
Citrus notes	Citral
	Geranyl nitrile
	Agrunitile
Spicy notes	
	Eugenol
	Clove
	Cinnamon
	Pepper
Carbinols and their esters	
Rosatol	Phenylethyl dimethyl carbinol
	Dimethyl benzyl carbinyl acetate
	Dimethyl benzyl carbinyl butyrate
Honey notes	
Ethyl phenylacetate	Methyl phenylacetate
	Citronellyl phenylacetate
	Eugenyl phenylacetate
	Phenylethyl phenylacetate
	Phenylacetic acid
	Cire d'abeille absolute
Additional base notes	
	Diphenyl oxide
	Cassione
	Frambinone
	Musk T
	Ambrettolide
Naturals	
Geranium	Camomile bleue
	Camomile Roman
	Palmarosa
	Carrot seed
	Guaiacwood
	Sandalwood
	Iris concrete
	Mimosa absolute
	Benzoin Siam
Blenders	
	Nerolidol
	Farnesol

JASMIN—COMPOUNDING NOTES

Basic Formula 1 (Classic)	Basic Formula 2 (Absolute Type)	Selected Variants and Modifiers
Benzyl acetate		
Benzyl acetate	Benzyl acetate	Benzyl propionate
		Benzyl valerianate
		Benzyl isobutyrate
		Dimethyl benzyl acetate
Phenylethyl alcohol and esters		
Phenylethyl alcohol		Phenoxyethyl alcohol
Phenylethyl acetate		Phenylethyl butyrate
		Phenoxyethyl isobutyrate
Hydroxycitronellal		
Hydroxycitronellal		Lilial
		Lyral
Linalool		
Linalool	Linalool	Linalyl acetate
Floral modifiers		
Geraniol	Geraniol	Citronellyl lactone
	DMBCA	
		Ionone alpha
		Isobutylbenzyl carbonate
	Eugenol	Isoeugenol
Green notes		
	cis-3-Hexenyl acetate	
	Violet leaf absolute	
	cis-3-Hexenyl benzoate	
Floral cinnamic derivatives		
Amyl cinnamic aldehyde	Hexyl cinnamic aldehyde	Cinnamic alcohol
		Benzyl cinnamate
Other jasmin chemicals		
	Hedione	Dragojasimia
		cis-Jasmone
		Dihydro jasmone

Basic Formula 1 (Classic)	Basic Formula 2 (Absolute Type)	Selected Variants and Modifiers
Anthranilates and Schiff bases		
Aurantiol	Methyl anthranilate	Other Schiff bases
Indolic		
Indol	Indol	Scatol
		Indolal
		Indolene
		Lactoscatone
Paracresol and esters		
Paracresol	Paracresyl acetate	
	Paracresyl phenylacetate	
Fruity		
Aldehyde C14		Aldehyde C18
		Aldehyde C16
		Gammadecalactone
	Maltol	Frambinone
		Cycloamylone
Naturals		
Ylang extra		Cananga
	Roman camomile	Foin absolute
	Celery seed	Immortelle absolute
		Mimosa absolute
		Benzoin siam
Blenders		
Benzyl alcohol	Benzyl alcohol	Nerolidol
	Benzyl benzoate	Isophytol

MUGUET—COMPOUNDING NOTES

Basic Formula	Selected Variants and Modifiers
Hydroxycitronellal	
Hydroxycitronellal	Lyral
	Lilial
Additional aldehydic notes	
Cyclamen aldehyde	Muguet aldehyde
	Dupical
	Aldehyde C7
	Aldehyde C10
	Oncidal
Linalool	
Linalool	Ethyl linalool
	Rosewood
	Tetrahydrolinalool
	Dimetol
	Linalyl acetate
Phenylethyl alcohol	
Phenylethyl alcohol	Phenylethyl methylethyl carbinol
	Phenylethyl isobutyrate
	Phenylethyl formate
	Phenylethyl phenylacetate
Rose alcohols	
Citronellol	Rhodinol
	Geraniol
	Citronellyl acetate
Floral modifiers	
Benzyl acetate	Hedione
	Ionone alpha
	Isoeugenol
Green notes	
Phenylacetic aldehyde	Phenylacetaldehyde dimethyl acetal
	cis-3-Hexanol and esters
	Hivertal
	Phenylacetaldehyde glyceroacetal
	Acetaldehyde diphenylethyl acetal
Floral cinnamic derivatives	
Hexyl cinnamic aldehyde	Amyl cinnamic aldehyde
	Linalyl cinnamate
	Cinnamic alcohol
	Cinnamyl acetate
	Phenylpropyl alcohol
	Phenylpropyl acetate

Basic Formula	Selected Variants and Modifiers
Sweet floral	
	Heliotropin
Indolic	
Indol	Indolal
	Civet
Naturals	
	Sandalwood
	Jasmin absolute
	Rose absolute
Blenders	
Benzyl benzoate	Farnesol

LILAC—COMPOUNDING NOTES

Basic Formula	Selected Variants and Modifiers
Phenylethyl alcohol	
Phenylethyl alcohol	Phenoxyethyl isobutyrate
	Phenylethyl formate
Terpineol	
	Terpineol
Ylang	
Ylang extra	Benzyl acetate
	Linalool
	Methyl paracresol
	Methyl benzoate
Muguet notes	
Hydroxycitronellal	Lilial
	Lyral
	Cyclamen aldehyde
	Citronellol
	Geraniol
Green notes	
Phenylacetaldehyde	Phenylacetaldehyde dimethyl acetal
	Syringa aldehyde
	para-Tolyl aldehyde
	Methyl acetophenone
	Cuminic aldehyde
Anisic notes	
Anisaldehyde	Anisic alcohol
	Anisyl acetate
	Acetanisol
Sweet floral	
Heliotropin	Vanillin
Floral cinnamic derivatives	
Cinnamic alcohol	Phenylpropyl alcohol
	Phenylpropyl acetate
	Amyl cinnamic aldehyde
Isoeugenol	
Isoeugenol	Methyl isoeugenol
Indolic	
Indol	
Naturals	
	Cinnamon bark
	Petitgrain
	Styrax oil
Blenders	
Benzyl alcohol	

CARNATION—COMPOUNDING NOTES

Basic Formula	Selected Variants and Modifiers
Ylang	
Ylang extra	Ylang no. 2
	Cananga
	Benzyl acetate
Rose notes	
Geraniol	Citronellol
	beta Damascone
	Rose oxide
	Geranium
	Citronellyl acetate
	Phenylethyl alcohol
	Orthomethoxy benzyl ethyl ether
Floral modifiers	
	Linalool
	Isobutyl benzoate
	Hedione
Spicy notes	
Eugenol	Isoeugenol
	Methyl eugenol
	Isoeugenyl acetate
	Benzyl isoeugenol
Honey	
	Eugenyl phenylacetate
	Isobutyl phenylacetate
	Phenylethyl phenylacetate
Floral cinnamic derivatives	
Cinnamic alcohol	Phenylpropyl alcohol
	Cinnamyl acetate
	Methyl cinnamate
	Amyl cinnamic aldehyde
	Hexyl cinnamic aldehyde
Salicylate	
Benzyl salicylate	Amyl salicylate
	Hexyl salicylate
	cis-3-Hexenyl salicylate
Vanilla-sweet	
Vanillin	Ethyl vanillin
Heliotropin	
Naturals	
	Clove, Carrot seed oil
	Pepper, Guaiacwood
	Rose oil, Cedarwood
	Iris concrete, Citronella

HYACINTH—COMPOUNDING NOTES

Basic Formula	Selected Variants
Phenylethyl alcohol	
Phenylethyl alcohol	Phenylethyl acetate
	Phenylethyl formate
Benzyl acetate	
Benzyl acetate	Benzyl propionate
	Benzyl formate
Green notes	
Galbanum oil	Galbanum resin
Hydratropic aldehyde dimethyl acetal	Hydratropic aldehyde
	Phenylacetaldehyde
	Phenylacetaldehyde dimethyl acetal
	Phenylacetaldehyde glyceroacetal
	Vertral
	cis-3-Hexanol, acetate
Floral modifiers	
	Linalool
	Cyclamen aldehyde
	Lilial
Rosatol	Heliotropin
Floral cinnamic derivatives	
Cinnamic alcohol	Phenylethyl cinnamate
Phenylpropyl alcohol	Styrax
	Amyl cinnamic aldehyde
Salicylate	
Amyl salicylate	Phenylethyl salicylate
	Benzyl salicylate
Eugenol	
Eugenol	Methyl eugenol
Indolic	
Indol	Indolene
	Indolal

VIOLET—COMPOUNDING NOTES

Basic Formula	Selected Variants and Modifiers
Ionones	
Methyl ionone	
Ionone alpha	Ionone beta
Phenylethyl alcohol and esters	
Phenylethyl alcohol	Phenoxyethyl propionate
	Phenoxyethyl isobutyrate
Green notes	
Methyl octine carbonate	Methyl heptine carbonate
	cis-3-Hexenyl methyl carbonate
	Nonadienal
	Parmavert
	Violet leaf absolute
	Liffarome
	Phenylacetaldehyde glyceroacetal
	Fiorivert
Anisic	
Anisaldehyde	
Jasmin notes	
Amyl cinnamic aldehyde	
	Dragojasimia
	Isojasmone
	Benzyl phenylacetate
Sweet floral	
Heliotropin	Cassione
	Frambinone
Woody	
	Iso E super
Naturals	
Sandalwood	Cedarwood
	Geranium
	Rose
	Cassie
	Iris
Blenders	
Benzyl benzoate	

NEROLI-ORANGE BLOSSOM—COMPOUNDING NOTES

Basic Formula	Selected Variants and Modifiers
Phenylethyl alcohol	
Phenylethyl alcohol	
Linalool	
Linalool	Ethyl linalool
	Dimetol
Fresh	
Petitgrain	Sweet orange
Linalyl acetate	Ethyl linalyl acetate
Green notes	
Acetophenone	Cortexal
	Hivertal
Rose alcohols and esters	
Geraniol	Nerol
Geranyl acetate	Neryl acetate
	Geranyl acetone
	Citronellol
Anthranilates and Schiff bases	
Methyl anthranilate	
Aurantiol	Meaverte
Orange flower chemicals	
Methyl napthyl ketone	Nerolin bromelia
	Nerolin yara yara
Honey notes	
Phenylethyl phenylacetate	Linalyl phenylacetate
	Benzyl phenylacetate
	Phenylacetic acid
Aldehydic notes	
Aldehyde C10	Aldehyde C8
	Alcohol C12
Indolic	
Indol	Indolene
Blenders	
Benzyl alcohol	Nerolidol

7

The Technique of Matching

When student perfumers have familiarized themselves with the basic raw materials and mastered some of the most important floral accords, they may well feel that they are ready to start creating new perfumes of their own. There is certainly no harm in occasionally giving their creative impulses free reign, and perhaps, through luck or talent, they will discover interesting novel accords even at this early stage. However, although this can be an exciting and highly motivating experience, it will usually end in frustration when what initially seems to be an interesting idea fails to take on the form of a well-balanced and aesthetically pleasing perfume (despite the enthusiastic comments from well-meaning friends and relations to whom the first results are presented). Having learned that the creation of a new fragrance is far more difficult than they had imagined, they will now be ready to embark upon one of the most important parts of their early training: the imitation or matching of the great perfumes of the past and present.

MATCHING AS A METHOD OF LEARNING

Learning from the work of the great masters is important in the study of all the arts. This is the reason why art students copy paintings in museums and why the young Richard Wagner copied, note for note, the scores of Beethoven's symphonies. The aesthetic laws of perfumery, the laws of harmony and contrast, of unity of impression and memorability, of depth and impact, cannot be fully expressed in words, but they are embodied in the great perfumes. The best way to learn these laws is to recreate, that is, to match these perfumes.

Until a few years ago the matching of a perfume was a tedious and often frustrating exercise involving a long series of painstaking trials, usually ending up with a not very good likeness of the original—although occasionally with a quite passable new perfume. Even with the help of senior colleagues it could take many years to acquire a working knowledge of the most basic repertoire of fragrance types. Today, however, with the availability of detailed analytical information from gas chromatography (GC), it is possible for students to learn much more quickly and to achieve results that were unobtainable by an earlier generation of perfumers. Herein, however, also lies a danger: Too much information can frequently be as harmful to the development of student perfumers as too little can be frustrating. Simply to give students a number of formulas does not turn them into masters of their profession. The hard work of matching is an essential part of the training process, not only in teaching the important skill of olfactory analysis but also in being one of the best ways of learning, through continual trial and error, the ways in which materials work together in various combinations.

Most student perfumers today work within reach of GC information. But it should be part of the job of the senior perfumer responsible for their training to ration out the information in such a way as to ensure that they learn the necessary skills of olfactory analysis, learning to recognize the accords that make up the structure of a perfume and discovering for themselves the level at which each material works best in the final composition. At the same time the judicious giving of information will allow students to make sufficiently rapid progress to maintain their interest and excitement.

While studying in this way students must at all times be encouraged to have an intellectual grasp of the formula on which they are working by trying to enter into the mind of its original creator. A student who finishes the work with little understanding of the way in which the perfumes are put together will never go on to create successful perfumes of his own.

In working on a perfume students should be constantly looking for the important accords within the formula and the often surprising effects of these accords. Students may find, for example, that a certain combination of ionone, benzyl acetate, and vanillin produces a raspberry like effect in a perfume that one might at first expect to come from a single chemical. Students should look for similarities between structures of different perfumes and learn to recognize the basic accords that make up the major families. But above all they must maintain a continual curiosity toward the perfumes on which they are working, trying not only to arrive at as close a match as possible but also to understand the significance of each material found to be present.

A final reason why mastering the technique of matching is essential is because there is a form of imitation involved in true creation. Whenever perfumers write down or modify a formula with the intention of achieving a specific imagined effect, they "match" their mental image of what they want to achieve. The skills that they have learned in matching the work of other perfumers enables them to realize their own creative ideas.

Although gas chromatography has become indispensable in the matching of perfumes, we will begin by considering the techniques of matching purely as they relate to the perfumer's most valuable piece of analytical equipment—the one that needs the most practice—the human nose.

THE ANALYTICAL SMELLING TECHNIQUE

Everything that was said in Chapter 3 about the general techniques of smelling applies also to the analytical smelling of mixtures. The suggestion, to observe the odor of a new material at its various stages of evaporation, acquires new importance in the case of blends. Due to their different vapor pressures, mixture components evaporate at different rates and hence make themselves felt to different degrees at the various stages of evaporation. The most volatile components are most readily apparent during the early stages of evaporation, while the most tenacious components of the blend can best be observed at the final stages, when all the other components have faded away. Studying the dry-out note is particularly useful, since many of its components, such as the musks or benzyl salicylate, may be relatively difficult to detect in the top note of the perfume.

Smelling with total concentration is, if anything, even more important when analyzing blends than when studying single materials. In time students will naturally acquire the ability to scan a composition focusing on different components in sequence. They will learn to ask themselves directed questions such as Does the perfume contain salicylate? or Does the spicy character come from one of the sweet spices containing eugenol, such as clove or pimento, or could it be eugenol itself in combination with some other material? Focused smelling also entails the ability to put aside mentally a component that one has clearly recognized and smell around it to detect others. A trick, suggested by Paul Jellinek, is to saturate your nose with a component that is definitely present to make it more sensitive to the other ones. This principle is important also as a match comes closer to the perfume being copied. Materials that we may have failed to detect at an earlier stage begin to show up by contrasting the incomplete copy with the original. Another such technique is to compare two smelling strips of the same perfume at two different stages of its evaporation to make the components that have faded on the older smelling strip stand out more clearly on the fresher one.

Roudnitska recommends, as an aid to analytical smelling, that a drop of the perfume to be matched be allowed to fall on a smelling paper wide enough for the solution to spread without reaching the edges. This results in a degree of physical separation of the components, as in paper chromatography, a principle which was used as the basis for gas chromatography.

SMELLING IN PATTERNS

Even when smelling a complicated blend of materials experienced perfumers are readily able to recognize the presence of certain natural ingredients. They do not need to think of the hundreds of possible individual components each may contain. The specific combinations form an olfactory pattern that perfumers are immediately able to identify as, for instance, patchouli, geranium, or tolu balsam.

This ability to "smell in patterns" can also be applied to the smelling of complete perfumes. Perfumers learn to recognize the patterns that are characteristic of the classic perfume types. On first smelling a new men's fragrance of the Fougere family, for example, a perfumer does not first perceive lavender and patchouli, oakmoss and coumarin, and then say: "Aha, that's a Fougere." To the perfumer, the order is reversed. He or she recognizes the Fougere pattern and then may say, "Since this is a Fougere, it should contain lavender, etc.," and smelling into it further, the perfumer looks for these components. The perfumer then focuses attention on those features that make it stand out from the conventional Fougere pattern, and mentally "setting aside" the customary components, he or she will detect nuances such as an anisic note or a fruity green character.

As will be discussed in a later chapter, most perfumes fall into well-defined and gradually evolving families. Recognizing the group to which a perfume belongs, and its relation to other perfumes within the group, will immediately give the well-trained perfumer an idea of much of the formula.

The ability to perceive in patterns is something that the human brain (and indeed the brains of most animals) is particularly good at. We can instantly pick out individual words in a room full of chattering people, or listen to a solo violin amid a complete orchestra. Our ability to read and to identify everyday objects from among a mass of incoming visual information is also based on pattern recognition. This is something that the brain continues to do much better than even the smartest computer.

Like all skills pattern recognition can be greatly enhanced by practice. Specialized areas of perception are what distinguishes the expert from the ordinary person, whether it be in the game of chess, in medical histology, in perfumery and GC analysis, or in any other field. Because smelling in patterns is an essential part of the perfumer's skill, the perfumer's training centers around complexes, both those which occur in nature and those created by perfumers.

MATCHING BY NOSE

As has already been mentioned, much of the work of matching is nowadays carried out in conjunction with gas chromatography. But for the moment let us forget that the GC exists and continue with our discussion of matching using only the nose as a guide.

Let us return to the sample to be matched. After having smelled it analytically and related it to other perfumes who structure is already known, we may begin to make a blend, in proportions that seem reasonable, of those components that have been recognized. This is done cautiously. "If in doubt, leave it out" is a good precept, for it is much harder to remove a wrong component from an attempted match than it is to add a missing one.

Because of the trouble that may result in having the "wrong" components in a match, it is generally better to adopt a "linear" technique of composition, that is, to work with single-aroma chemicals and naturals rather than with bases. A classical, widely used base may, however, be used when clearly identified by its characteristic olfactory pattern within the overall perfume. Bases may also be used to give the finishing touch to a nearly completed match or when the aim is not so much a close match as a composition "inspired by" the original.

The art of matching also entails the skill of comparative smelling. Again and again, the perfume to be matched is compared with the samples that represent the successive attempts at matching. Although at first it is best to compare only two samples—the original and the latest match—it is always important to go back to earlier ones to make sure that we haven't strayed somewhere along the way. To be totally honest with ourselves when asking the question "which of these experimental blends is closest to the original" is often difficult; we always prefer to think that the latest attempt is the best. This is particularly true at the end of the day or on a Friday evening when we like to go home feeling that real progress has been made, only to discover in the cold light of the following day or week that we still prefer an earlier trial.

One of the difficulties in comparative smelling stems from the fact that the sample that is smelled second is strongly affected by our perception of the first, especially if the two are smelled in quick succession. Adaptation causes the first sample to appear stronger than the second. In addition a carry-over effect may occur in which a clearly perceived note in the first appears also to be present in the second, even if this is not the case. To minimize such effects, it is best not to dwell too long on either sample before going on to the other. However, the pause must not be too long since the recall of complex notes quickly loses its sharpness.

Once you think you have identified a distinct point of difference between the two samples (e.g., the original contains patchouli, while the match does not), test your hypothesis by switching the two smelling strips between the hands inattentively, until you no longer know which is which. Now try to identify the original using only your new hypothesis as a guide. If you find this easy, your hypothesis was probably correct.

Another technique that is useful in testing a hypothesis is to hold a smelling strip of the material that you think should be added close to the one with the trial match, in such a way that you smell the two simultaneously. This amounts to a tentative (though imprecise) addition of the component considered.

Now compare the samples again, concentrating this time on other aspects of the fragrance such as the nature of the floral accord, the aldehydes, or the animal notes. To take a complete inventory of a complex perfume and its match may take considerable time and effort. It involves comparing the samples over and over again, at all stages of evaporation, from the first top note to the dry-out.

Just how far to carry on the comparison between the original and the match before making up a further trial is a question of judgment and experience. There is always the temptation to rush ahead, adding ever more ingredients usually without stopping to consider those that should be discarded. In so doing, the formula may become overcomplicated, making it impossible to sort out the good ideas from the bad. A golden rule in matching (as in creation) is to repeatedly go back to the basic structure of the perfume, simplifying wherever possible before proceeding with the decorative nuances that surround it.

"TUNNEL SMELLING"

One of the difficulties in matching is the tendency to be, at first, overly pleased with what one has just made. In smelling our new experimental blend, we are apt to find it marvelously close to the original. But there is often a certain amount of "wishful smelling" involved in this phenomenon. We smell what we expect and hope to smell. In other words, what is at work is something that we might call "tunnel smelling."

In smelling a complex fragrance, we do not always grasp it in its full complexity. Our perception is dominated by those features that strike us most powerfully, and if we manage to match these features reasonably closely, we may feel that we have come a long way toward matching the perfume. Indeed we may have done so, but someone else whose attention has been struck by other features of the original, which we may have more or less overlooked, may immediately detect discrepancies and be far less impressed by the closeness of our match.

Simply letting time pass and coming back to the perfume on the following day helps a great deal in overcoming tunnel smelling. Taking samples home to smell is also useful, since we tend to smell differently in different environments. A car often provides an excellent environment for smelling, though we would not recommend smelling on the way home from work with one hand on the steering wheel and the other holding a number of smelling strips!

An obvious way to overcome the subjectivity of tunnel smelling is to ask colleagues for their opinions. Whether this is helpful depends on the personalities of the colleagues whose advice we seek. Some people seem constitutionally unable not to find anything wrong, and we may go away merely discouraged by their dismissive remarks. The student learns quickly whom to go to for advice. Here the role of the sympathetic teacher is of great importance in providing constructive criticism and encouragement. Among senior perfumers, or between groups of students, much will depend on the working atmosphere within the department. Where each perfumer is judged by his or her individual success, a spirit of competition is fostered, and constructive criticism of each other's work can become the exception rather than the rule. For this reason students should be encouraged from the beginning to learn to work together in friendly cooperation, sometimes working in pairs, as well as in competition.

THE USES OF GAS CHROMATOGRAPHY

The gas chromatography has been called a "super blotter," or "smelling strip." The smelling strip separates a blend, over a fairly long time span, roughly into groups of the most volatile, less volatile, and most tenacious components. The gas chromatograph accomplishes the same not only more quickly (the entire analysis of a perfume normally takes about 90 minutes) but also far more thoroughly. In the hands of a skillful operator, the perfume can be nearly completely separated into its individual components, which leave the exit port one after another.

Many excellent books have been written on the theory of gas chromatography so we will not discuss the subject in detail in the present context. Briefly it is based on the differences in the energy required for each component of a mixture to pass from the surface of a liquid phase, onto which it has been adsorbed, back into the vapor phase.

This rather complicated idea can be better understood by looking at what actually happens in a gas chromatograph. The sample to be analyzed is vaporized as it is injected into a stream of inert carrier gas, the mobile phase. The gas moves through the heated column, where it passes over the surface of the stationary phase, which consists of a liquid adsorbed onto the surface of an inert carrier. The individual components are separated as they pass through the column owing to the different speeds at which they are repeatedly adsorbed and desorbed across the surface of the stationary phase. The components are each detected electronically as they leave the exit port and recorded as a series of peaks on a moving chart, the chromatogram. The height of an individual peak, or more accurately the area beneath it, is roughly proportional to the amount of the material being recorded. The characteristic time taken for a component to go through the column, under given conditions such as temperature and the flow rate of the carrier gas, is called its "retention time". In addition to the detection of the components by electronic means, they may be smelled and identified by a perfumer as they leave the exit port. This requires a special kind of experience because materials can be perceived as having rather different odors when emerging in a stream of hot gas than when smelled on a smelling strip.

In the early days of gas chromatography glass columns were used, packed with granules of the stationary phase material. For most purposes these have now been replaced by capillary columns, often up to 60 meters in length, which are coated on the inside with the stationary phase material. These columns allow the injection of a much smaller amount of the mixture to be analyzed and achieve very much more efficient separation. Materials can be detected that represent less than 0.01% of the total mixture.

A number of different materials can be used to act as the stationary phase. The most widely used materials in perfumery analysis are the so-called polar materials (carbowaxes or their equivalents). Polar columns separate the components in much the same way as the perfumer is used to smelling them as they evaporate from a smelling strip. By using two different types of columns—one polar and one nonpolar—it is possible to obtain more information than from just one. Sometimes two or more materials will, under certain conditions, have identical retention times, resulting in "hidden peaks" on the chromatogram. By using the different types of columns or by changing the temperature programming, these components can be made to separate on the chromatogram. Nonpolar columns can often be run at higher temperatures than polar columns, making it possible to identify some of the higher boiling materials, such as Schiff bases, which do not appear when polar columns are used.

Besides the traditional type of chart recorder, chart integrators, or "printer-plotters" as they are often called, may be used. Printer-plotters automatically calculate the percentage of each component in the total mixture and print this out in the form of a list of retention times and percentages. But, unless a correction factor is applied to each component, these values, which are based on the characteristic response of the detector, are not normally accurate to within more than a 10–20% variation.

As a further development, the GC machine can now be connected to a mass spectrometer (MS) and a computer, which are able to give a positive identification of the various components. In effect the perfumer can be given a list of all identified materials, often comprising some 95% of the compound, together with their approximate percentages. Often a hundred or more materials will be identified in this way. Since any natural materials are broken down into their individual components, along with the rest of the perfume, many of these ingredients will be included in the printout.

To convert all this information into the form of a perfume formula requires considerable experience on the part of the GC analyst and perfumer. Some perfumers learn to specialize in this work, combining their perfumery skills with a knowledge of GC technique. Such a perfumer working alongside the GC/MS is often able to produce results unobtainable by a straight reading of the analytical printout.

The first step in the reconstruction of the original fragrance is to produce a trial formula based on all the main components given in the analysis. The experienced perfumer will also be able to spot the presence of specific natural materials by the characteristic patterns of peaks that they produce on the chromatogram, and will know at a glance the approximate percentage of the compound that these represent. The "fingerprints" of certain materials such as geranium, patchouli, armoise, sandalwood, and ylang and even their different qualities, are more easily recognized by the experienced human eye than by reading through the analytical information, coming from the integrator, or by the use of a computer. (This is another example of the ability to perceive patterns discussed earlier.) Even amounts of such natural products down to as little as 0.1% can easily be seen on a well-produced chromatogram. The presence of other more "difficult" naturals may be suggested by the MS identification of one or more highly characteristic materials, for example, as by the presence of myristicene in nutmeg oil or of styrene in styrax. However, some materials, including naturals such as galbanum and juniper berry, and trace amounts of very strong materials, are most easily detected and identified olfactorily by smelling their components as they leave the exit port of the GC machine.

The first trial compound, containing all the major ingredients that have been positively identified, is then compared to the original, both by smelling and by further GC analysis. A comparison of the two chromatograms, that of the original perfume and the first trial copy, will give an immediate indication of adjustments to be made and gaps to be filled. The perfumer then will use his or her skills in analytical smelling to suggest the presence of other materials that show up more clearly by contrasting the two samples. Confirmation of these materials may be searched for in the analysis, or among the components smelled as they come off the end of the column.

The perfumer will also look for combinations of materials, often in trace amounts, that suggest the presence of bases within the composition. These can often be reconstructed from within the analysis and compounded separately rather than incorporated individually into the formula. Well-known bases can be recognized by the presence of their major constituents in the characteristic proportions. For example, the presence together of phenoxyethyl isobutyrate and dimethylbenzyl carbinyl butyrate would suggest a particular type of fruit base. Here the perfumer would try to read the mind of the original creator of the perfume, thinking of how the structure of the perfume could have been put together rather than just of its individual components.

Several trial compounds may have to be made before a reasonably close analytical match is obtained. One of the problems that arises at this stage is sorting out the contribution that various components can make to the total quantity of a single ingredient found in the analysis. If citronellol is present, it may be because it occurs as a specified component in the formula, because it comes from a number of different essential oils, or because it is included in one or more bases. Many of the terpenes, such as *d*-limonene, are present in numerous essential oils, any number of which may be present. Trying to get back to the original formula requires considerable deductive powers and ingenuity on the part of the perfumer.

It is advisable when trying to match a perfume to obtain as fresh a sample as possible, since an older sample may have undergone chemical changes that make the analysis more difficult. Schiff bases may have formed in the presence of methyl anthranilate, and in alcoholic fragrances aldehydes will have been progressively converted to their diethyl acetals. Oxidation of some of the terpenes may have taken place, changing their relative proportions.

From what has been said so far, it might be assumed that the GC/MS provides the complete answer to the task of matching. Occasionally indeed a perfume may, if sufficiently simple in structure, be reconstructed quite accurately in this way with little recourse to the human nose. But more often than not, even a duplication that closely follows the analysis of the original will require some additional work by the perfumer. Many natural materials are difficult to detect when present in very small amounts, yet in combination they can add enormously to the richness of the perfume. Resinous materials, owing to their high boiling points, may not always pass through the column, although some of their constituent ingredients may give an indication of their presence. Such materials can only be found by analytical smelling of the original perfume. Sometimes a material unknown to the MS data bank will be found. This material may be smelled as it leaves the exit port, and if possible a replacement would be used to obtain a similar effect in the end product.

So far we have considered only the analysis of either perfume compounds or perfumes in alcoholic solution that may be injected directly into the GC. Frequently, however, the perfumer will be asked to match the perfume of a functional product such as a soap or fabric conditioner. Most companies have now developed techniques for the extraction of perfume ingredients from most types of bases. However, some degree of selectivity can take place in the extraction process, and the resultant compound may be quantitatively different from the original. In such cases perfumers need to rely rather more upon their own analytical smelling, using the GC/MS analysis as a guide rather than as the source of precise information.

Although perfumers will sometimes try to produce as exact a duplication of a perfume as possible, and much can be learned by doing so, the most common application of the GC/MS-plus-perfumer analyses has been for the rapid development of approximate matches. The formulations of these matches are then used for the creative development of modifications, adaptations for different media, and so on, or even as the inspiration for genuinely original creations.

Some have viewed the rise of the GC/MS-plus-perfumer approach as a threat to the traditional role of perfumers in the sense that it makes their function obsolete. This view is unfounded, since GC/MS analysis is a technique of matching, not for creation. Nevertheless, this technique is in the process of revolutionizing the perfumery industry. It is making matching, once an important part of the daily work of most perfumers, the province of small teams of specialists. Consequently it both enables and forces perfumers to concentrate their efforts on creation.

GC/MS analysis is contributing to the piercing of the veil of secrecy that in former days surrounded perfume formulations. This opening up of information has also affected the economic structure of the industry. The revolution continues.

BRIEFINGS INVOLVING MATCHES

When a client requests a match of an existing perfume, he or she may be pursuing one of three goals.

Having noticed the continuing success of a certain type of perfume on the market, or one that is becoming trendy, the client may wish the intended product to take advantage of this success. In such a case what the client is looking for are proposals "in the direction of" or "in the family of" the type indicated.

The situation is different if the client, having noticed that a certain product is successful in the marketplace, intends to launch a product that is perceived by the public as being very similar to the successful product. In this case the client selects a product appearance, product name, and a package design that are as close to the model as ethics or regulations permit, and looks for the same degree of closeness in the fragrance.

A different situation again prevails if the client already has a product on the market and decides, for any one of a number of reasons (a change in import regulations, unhappiness with the current supplier, economy, etc.) to replace its fragrance without that change being noticed by the product's users.

"In the Family" Matches

The practice of composing new fragrances that lie "within the family" of well-known existing ones is an integral part of the culture of perfumery. It is at the root both of fragrance trends and of the fact that we can speak, within a given market and a given time frame, of typical shampoo notes, typical household cleaner odors, and so on.

To create perfumes that are "within the family" of given models, the perfumer need not have a precise imitation of the model but must be familiar with its basic characteristic accord. Actually the perfumer must have such knowledge of the leading perfumes within a market even if the objective is to create a distinctly new perfume. Consumers do not usually like perfumes that depart too far from what they have come to regard as normal within a given category. Successful revolutions are as rare in perfumery as in other applied arts.

Close Matches

Calls for matches that go beyond family resemblance and aim at closeness to a specific model entail technical, practical, and ethical problems. The technical problem lies in the fact that even with the help of GC/MS, far more effort is involved in creating a close rather than an approximate match. The difficulties are multiplied if the character of the perfume is considerably modified by the components of the product base and if, as is usually the case, the cost limits are well below the cost of the original fragrance.

The practical problem stems from the fact that briefings calling for close matches are often based on a fundamental misconception. In thinking that they can capture the model's success for their own product by closely matching the model's physical features and perfume, the manufacturers and marketers of imitative products overestimate the effect of these features on the overall acceptance of the product and underestimate the importance of such features as package, distribution, and pricing. This is particularly the case in alcoholic perfumery, but it holds also for all other product categories.

The hopes of the producers of imitative products are therefore often disappointed. Where such products are successful in the marketplace, it is not because of the closeness of their perfume to that of the original but simply because the product offers good value for money. The perfumer's time would have been better spent, and the client's interest better served, had the perfumer expended time and effort in making a perfume "in the family of" that delivers optimal performance and optimal value for money in the customer's product base.

The situation is different in cases where the client aims at closely imitating not just a model product's physical features but also its packaging in all aspects, including the product name. We are then in the realm of counterfeit products. There is no need to expand upon the ethical and legal problems associated with being an accomplice to the manufacture of products of this kind.

A special category of imitative products are the so-called copycat perfumes. Here no attempt is made to imitate the model's name or package. The selling proposition is: Just as nice a fragrance, at a much lower price. The ethical problems involved here are subtle and open to argument. It is established practice among most reputable suppliers that the supplier who has created the original will not, using his or her knowledge of the formula, create copycat versions. The position here has become less well-defined recently, owing to the ability of most companies with GC/MS to make close countertypes of successful products coming on to the market. The supplier of the original perfume may well feel aggrieved if he or she is the only one unable to make further use of a personal creative idea. It becomes a question of judgment as to how close the perfumer may go to the original.

Replacement

What is demanded here is not just a fragrance which to the general public appears similar to the original. The match must not be recognized as being different by the most demanding judges of all, the product's regular users. Practical considerations, as well as the ethics of customer-supplier relationships, dictate that assignments of this type should, where possible, be given to the fragrance house that created the original.

The ethics of giving briefings of this type to other suppliers depend on the details of the situation. The case is straightforward if using the original supplier is no longer possible, for instance, because the supplier has gone out of business or because of newly imposed import restrictions. Where alternative suppliers are called upon in order to put pressure upon the original supplier, there are major problems, both ethical and practical.

The ethical problems touch upon the interests not of the marketer but of the supplier industry. The economic basis of this industry rests upon profit margins sufficient to recover, within a reasonable time period, the expenses incurred in the development of fragrances. The practice of looking for lower-priced replacements from other suppliers denies such margins to the original supplier.

The practical problems arise from the fact that briefings of this type are of questionable value to the alternative supplier. They demand a great deal of effort while offering, by their very nature, slim profit margins. Moreover the business gained by winning such a brief is usually short-lived because there is no natural limit to the customer's desire for lower costs.

Where briefings to alternative suppliers are prompted by dissatisfaction with the current supplier or by fear that the current supplier may not, in the future, be able to satisfy the customer's requirements (e.g., for reasons of limited production plant or uncertain access to raw materials), they fall in a gray area and should, in each case, be judged upon their individual merits.

Although it does not lie within the intentions of this book to promulgate ethical precepts for the fragrance industry, the perfumer should be aware that any personal esteem gained and even success in the profession depend to a considerable degree both upon his or her personal ethics and upon those of the company for which the perfumer works.

Part II

Aesthetics and the Fundamentals of Composition

8

The Biological Basis of Aesthetics

Although this book is primarily concerned with creative perfumery, in this chapter we will be looking briefly at the function of smell in nature and the implications that this can have for the work of the perfumer. Smell is the most primitive of the senses, and our response to certain odors is deeply embedded in the subconscious mind. Much of what we know to be true, from our experience as perfumers, can be related to the origins of our olfactory sense and the part that it has played in the evolution of our species.

Smell, or the process of olfaction, is a chemical sense. Unlike the senses of hearing and vision which are stimulated by energetic phenomena, it depends upon the detection of airborne molecules coming in contact with specialized chemoreceptive cells.

Although evolution has taken place over many millions of years, nature has been surprisingly conservative in the development of the olfactory mechanism. The basic type of receptor cell found in primitive animals has changed very little during the course of evolution, remaining much the same in groups of animals as diverse as insects, birds, fish, and mammals. What has evolved, however, is the ability of the brain to make use of the incoming information from the environment: through the innate recognition and preference for certain types of smell, and through the ability to learn and to associate them with successful behavior patterns such as the sourcing of food or the escape from danger.

In humans the olfactory receptor cells lie in the mucous membrane at the top of the air passages on either side of the nasal septum. They occupy a total area of about 2 cm, which is small compared with most other mammals. Evidence from both anatomy and embryology shows that the development of the olfactory tissue is closely linked to that of the pituitary gland which lies at the base of the brain. Among other functions the pituitary plays a key role in the coordination of sexual activity and reproduction. This ancient association between the sense of smell and the reproductive process is one that has important implications for work of the perfumer.

From the receptor cells nerves pass through the olfactory lobes at the front end of the brain direct to the central basal region, the part known as the "limbic system." This forms part of our deep-seated unconscious mind, being associated with the control of emotion and sexual activity, as well as with feelings of pleasure. In evolutionary terms it is also the oldest part of the brain, providing evidence of the early and continuing importance of the sense of smell in animal behavior.

Only after entering the limbic system does the olfactory message pass to the cerebral hemispheres, which form the major part of the human brain, where cognitive recognition occurs and the ability to associate the smell with its name. By this sequence of events our full awareness of a smell takes place only after the deepest parts of the unconscious have been activated. As a result, even without conscious recognition, smell can be the most evocative of our senses, linking us to past experiences and stirring our emotions at a level that we find hard to explain.

Compared with other mammals and even with our closest relatives, the apes, the sense of smell in humans is exceptionally poor. Nevertheless, it continues to play a significant role not only in our conscious lives but also, as we will see later, in our subconscious awareness.

If we define a smell as being any material or group of materials that is capable of being detected by the process of olfaction, then we may divide the smells that occur in nature into two groups: those that are produced by the general processes of the environment and those that are used as a form of communication between living organisms, both between individual animals, either of the same or of different species, and between plants and animals. The first group consists of simply chemicals that exist whether or not there is an animal with a sense of smell capable of detecting them; the second is the result of millions of years of natural selection and evolution. The results of evolution are there because they have a clear adaptive advantage to the organism involved. The smells that have evolved are not just random in their composition, they have evolved alongside the ability to smell, selected for their ability to stimulate the olfactory mechanism, and to convey a message to which there will be a behavioral response. They are nature's perfumes, the structure of which may be a guide to the perfumer.

The chemical sense as a means of communication between separate organisms first assumed biological importance with the evolution of sexual reproduction. Aquatic organisms, long before the development of sight or hearing, needed to locate and correctly identify members of their own species for pairing. This was achieved by the release of specific chemical materials by one individual which directly affected the behavior of another, resulting in the synchronized release of gametes and fertilization. Such chemical messengers, or pheromones, occur in almost all animals, from unicellular organisms to insects and higher mammals, and play an important part in the mating process along with visual and vocal displays. They have a direct influence on the nervous system of the individuals at which they are directed, affecting their behavior and leading to successful mating. They are either single materials or simple mixtures to which the animal's response is genetically predetermined. There is no process of learning involved. They are for the animal concerned quite literally "what turns it on."

Apart from chemicals that act directly as pheromones in the reproductive activity of higher animals, there are many other associated materials, sometimes coming from animal excrement, that cause an arousal of interest and act as innate recognition factors between animals within a species, or as a warning to others. Some of these, as we will see later, have an important function in perfumery.

This link between the sense of smell and reproductive activity not only goes back to the most primitive organisms but is also deeply embedded in the unconscious mind of all higher animals including humans. Research into human biology has shown that if the neural link between the nasal receptor cells and the pituitary is broken, sexual interest and function may be greatly impaired.

However, at some stage in our evolution the sense of smell became greatly diminished. This may have come about partly because of the adoption of an upright bipedal position which distanced the human nose from many of the sources of smell. At the same time pheromones ceased to be the irresistible sexual turn-ons that they were to our prehuman ancestors. In his book *The Scented Ape,* Michael Stoddart (1990) regards the desensitization of the olfactory system that occurred in humans as having played a key role in our evolution, protecting the pair-bonding between man and wife within a gregarious community and thus allowing for a long period of training for the young within a stable family situation.

Pheromones are still produced and undoubtedly play some part in our unconscious behavior patterns. It is possible, for example, that our "gut reaction" to the people we meet, whether we like them or not, has something to do with our subconscious awareness of the pheromones they produce.

Despite this switching off of the importance of pheromones other types of human odor continue to play an important role in the bonding between individuals: between mother and child, between man and woman, and between members of a family. Although our ability to recognize people by their odor is very much less than, say, that of a dog, it is clear that each of us has his or her own unique smell.

The olfactory desensitization that took place in our evolution coincided with the development of the higher brain, the neocortex. Behavior patterns and preferences, including those linked to our sense of smell, became increasingly learned by experience rather than being based on a circuitry of nerves hard-wired at birth.

Newborn humans have few olfactory preferences; most of the wiring has to be put in place during the early years of development, though we continue to learn and acquire odor preferences throughout our lives. The cultural implications of this and its relation to perfumery are of course immense, since each of us is able to learn and respond to an enormous range of odorous materials rather than only to those dictated by a genetically predetermined system of behavioral responses.

THE IMPORTANCE OF ANIMAL SMELLS IN PERFUMERY

Although human odor forms a natural part of our social environment, it has become a fundamental principle of civilized society that human beings should not smell of human beings. Much of this has come about through social development and the conditioning that we receive as children, based on hygiene and sexual modesty. But the neural pathways connecting the sense of smell to our deepest centers of pleasure remain, and humans have from the earliest times delighted in the use of perfumes rich in products of animal origin—of any animal, that is, except the human being. In ancient times, and through to the Middle Ages, natural civet and musk were used in their own right as perfumes, employed not only for their supposedly aphrodisiac properties but also to cover the even less desirable smell of their wearers' own unwashed bodies! Today these materials would be rejected for their animalic character, unless incorporated into a perfume at levels where the mental associations induced by them remain in the unconscious mind. Intensely animalic materials of natural origin, as well as their synthetic counterparts, fecal notes such as indol and scatol, are widely used at acceptably low levels in perfume compositions of every type. Many plant products also have associations with animal notes and probably have an unconscious erotic effect. Costus has a smell associated with the sebacious glands of the hair, cumin has an intense smell of sweat, ambrette seed and angelica contain powerful musk like constituents. The recent introduction of cassis bourgeons into many modern perfumes may be connected to its resemblance to the sexually related odor of the male cat. The consumer must of course be quite unaware of this association.

However, there are also many synthetic materials, and materials of plant origin, that resemble natural animal odors such as musk and ambergris without having the intense fecal or urinous overtones that require them to be hidden within a composition at a level where their presence cannot be consciously recognized. These materials may be quite different chemically to the natural product, but to us their odor and effect are similar. Included in this category are the amber notes derived from labdanum and clary sage, as well as the numerous synthetic musk materials such as galaxolide and musk ketone.

The use of a great variety of musk and ambergris notes in the composition of perfumes is of great significance in perfumery. When smelled as part of a perfume they are capable of producing a sense of pleasure associated with ancient neural pathways without the perceiver being aware of the true nature of the stimulus. They form an important part of nearly every successful perfume.

THE OLFACTORY RELATIONSHIP BETWEEN PLANTS AND ANIMALS

Floral notes are also among the most widely used in perfumery. To the human psyche they are the most beautiful of smells and a constant source of inspiration to the perfumer.

The sense of smell in animals and their behavioral response to the presence of certain types of molecule has been exploited by plants in the evolution of their own reproductive processes. Many plants produce smells that mimic either the pheromones of insects or olfactory materials associated with animal excrement in order to attract insects to their flowers for the purpose of pollination. Indol, which is a characteristic ingredient of the feces of many animals, occurs widely as a component of many flower oils such as jasmin, orange blossom, muguet, and lilac. Although most of the animals that pollinate flowers are insects, some small animals including rodents and bats are similarly used by plants, attracted by the means of odors which to the animal concerned have a genuinely erotic association.

Even in fungi the exploitation of animals for the purpose of spore dispersal has produced some remarkable relationships. Pigs have for centuries been used in the search for truffles, toward which they have an insatiable curiosity. The fungus that produces its fruiting body underground is now known to contain substantial quantities of a pheromone produced by the male boars for attracting sows. By digging up the truffle and breaking it into pieces, the pig is actually aiding in the dispersal of its spores. The fact that we also regard the truffle as so great a delicacy is no doubt linked to our own unconscious awareness of its link with animal pheromones.

But many plants went on from this relationship based on the mimicking of single erogenic materials to evolve vastly more complicated fragrances in order to advertise their presence to animals such as bees and other insects. Flowers were evolved that offered a reward of food to a visiting insect in return for a service to the plant, that of transferring pollen from one flower to another. Plants are in effect using the wings of insects to carry pollen from one flower to another, providing them with fuel by way of nectar.

This coevolution between plants and animals for their mutual benefit has not only resulted in the complex and individual fragrances that they produce but in the wonderful color and structure of flowers themselves. We do not know whether insects such as bees can smell precisely the same range of materials as ourselves or whether their olfactory world differs from ours in the same way as does their vision. Bees are able to "see" ultraviolet, and the patterns of flowers that guide the bee toward the source of food appear quite different to the bee than they do to us. But, since we are able to smell most constituents of essential oils, it is probably fair to assume that our range of smell is quite similar. Perhaps bees have a much clearer idea of the smell of benzyl alcohol, which occurs in many flowers, than do most perfumers. How it would actually smell to them we cannot imagine any more than we can imagine the color of ultraviolet.

But let us consider the development of this relationship between insects and flowers in a little more detail, since there are some important implications to be found for the perfumer. Not all plants are pollinated by insects or other animals, and many species such as grasses rely on wind pollination. But once the distribution of pollen by insects had been established within the plant world the evolutionary pressure to achieve the maximum efficiency became enormous. Plants competed with each other in advertising their presence to insects by the evolution of more and more wonderfully specialized floral structures and scents; insects in their turn became increasingly specialized for the feeding on nectar and pollen. Not only were the plants dependent on insects, but insects became dependent on plants. Each had a vested interest in the survival of the other. The greatest efficiency was to be achieved by ensuring that an insect should feed only on one species of flower at a time, so avoiding the wasteful distribution of pollen to other flowers. Some insects developed an innate ability to memorize a particular odor as being associated with a ready supply of food. Bees returning to the hive are able to pass on to their coworkers information as to the direction and distance at which a source of nectar is to be found together with its associated scent. Programmed in this way, the bee will then visit only this particular type of flower.

What is interesting from the point of view of the perfumer is that this coevolution between animal and plant has resulted in the synthesis by plants of complex fragrances, often made up of many hundreds of materials, rather than being based on just one or two materials as in the case of pheromones.

Distantly related plants, such as rose, jasmin, and lavender have quite independently gone down this road of complexity, based on different groups of chemical constituents. We may conclude, therefore, that complexity of odor has evolved as being the most effective way of evoking a desired response from an animal with the ability to smell and the ability to memorize odor. What is remarkable is that the particular combinations of materials that individual flowers produce to make up their fragrance have, to our own sense of smell, an identity far greater than a random mixture of as many ill-assorted chemicals. Perhaps we may infer from this, in view of the similarity of our receptor cells, that the balance of materials is as important to the olfactory mechanism of the bee as it is to our own in producing a sense of identity and aesthetic pleasure.

If we look further into the chemical makeup of fragrances produced by flowers, taking rose as a typical example, we find that the oil is usually made up of perhaps no more than four of five materials that represent the main bulk of the product, with frequently hundreds of other materials making up the balance. If we mix together in the laboratory only those materials that make up the bulk of the product, although smelling of rose, the composition has little of the true character, strength, or aesthetic appeal of the flower itself. It is the combination of all the other materials, mostly in trace amounts, that results in the fragrance that is uniquely rose.

This balance between simplicity and complexity also plays an important part of the structure of a well-made perfume, and is a subject to which we will return in the next chapter.

9

The Structure of a Perfume

A perfume is a blend of odiforous materials, which is perceived as having its own unique and aesthetically appropriate identity. It is a carefully balanced blend based on a definite structure in which each material plays its part in achieving the overall fragrance. What it is not is just a mixture of pleasantly smelling materials.

Apart from having a well-defined identity a perfume must meet a number of technical requirements. It must be sufficiently strong, it must be diffusive (which is not quite the same thing), it must be persistent, and it must retain its essential character throughout its period of evaporation. A well-constructed fine fragrance will still be recognizable many hours after it is applied to the skin. Perfumes designed for functional products must have a degree of persistence appropriate to the use for which they are intended. They must also be chemically stable in the end product.

The technique by which this is achieved is an essential part of the perfumer's art, and it needs many years of dedicated work to arrive at the level of experience necessary to formulate perfumes that are not only original but also well made.

No two perfumers work in exactly the same way any more than do painters or musicians. There are styles in perfumery as there are in any other art, and the techniques that are used by perfumers can be seen to have changed enormously over the past hundred years, reflected in perfumes as different in their structure as Shalimar, L'Air du Temps, and Trésor. These perfumes can be seen to epitomize three different approaches to perfumery technique that are typical of the era in which they were created. They are as different in their own way as are the paintings of classical French art in the 17th century, the impressionist movement and cubism.

All art depends upon form and structure. The satisfaction which we get from looking at a great painting by Monet, or from listening to a symphony by Mozart is in large part due to the structure around which the artist or composer has arranged his ideas. Most of these structures have been discovered by intuition and experiment; we do not have to understand why they work, but only that they do. They are part of the tradition and experience that make up the history of an art form. Once a successful structure has been discovered it can provide the inspiration for an infinite number of variations and need not be the preserve only of its creator. The symphony was largely the creation of Haydn, but nearly all the great composers since his time have used it as a vehicle for their own creative work. Similarly perfumers continue to build their perfumes around the great classical forms, such as the chypre, floral aldehydic, and oriental accords (these are discussed in chapter 12), to produce what to the lay consumer are new and original fragrances. In recent years, as in many of the other arts, there has been a move away from many of these classical structures with the invention of new styles of formulation and technique.

Perfumery in the western world can be said to have begun with the invention of such products as Eau de Cologne, Hungary Water (a product based on herbs such as rosemary and thyme), and Lavender Water. These were made up of comparatively simple mixtures of distilled and expressed essential oils diluted in alcohol. To these were added, as "fixatives," tinctures of animal products such as musk, ambergris, and civet; balsamic materials such as benzoin and myrrhe; and sweet materials such as vanilla and tonka. The washings from floral pommades were also used. Toward the end of the nineteenth century the synthesis of chemical odorants, together with the technique of solvent extraction from plant material, saw the beginning of a gradually evolving sequence of styles and techniques that led to the world of perfumery as we know it today.

Initially, at the beginning of the present century, the traditional structure of the earlier perfumes was for the most part retained. "Fresh" natural materials such as bergamot and lemon, in combination with other essential oils, formed a large proportion of the composition, supported by animal and balsamic fixatives. To these were added the newly discovered synthetic and derived materials such as vanillin, coumarin, hydroxycitronellal, vetiveryl acetate, and methyl ionone, together with the newly developed floral absolutes. Such materials provided perfumers with a whole new palette of odorants with which to work, and they were to be the inspiration behind a new generation of perfumes. Shalimar survives as a wonderful example of a perfume formulated in this way.

Gradually synthetics, including the aliphatic aldehydes and the many floral notes created around the turn of the century, came to play an increasingly important role in perfumery, providing the main inspiration for the creation of new types of fragrance. Natural materials continued to be used but more as modifying notes and to provide richness to the composition. Working with defined chemical products allowed perfumers to study systematically the relation between materials of differing volatility in the building up of perfumes which were aesthetically satisfying and performed well in use. This approach led to the type of technique generally associated with the great perfumer Jean Carles. Carles constructed his perfumes on a basis of carefully selected empirical accords within a structure formulated around a base note, which included materials of low volatility, a middle or modifying note, which included materials of medium volatility, and a top note, which included the materials of the greatest volatility. Such perfumes as Canoe, Ma Griffe, L'Air du Temps, and Cabochard reflect this type of structure, the influence of which is seen in many surviving and recent perfumes. All of these perfumes have a wonderful transparency of texture, an almost three-dimensional quality that seems to lead us into the heart of the fragrance.

During the past 20 years there has been a gradual move away from this style of perfumery toward a new generation of perfumes based on a quite different type of structure, a style very much associated with the work of Sophia Grojsman. In this comparatively few materials, sometimes as few as four or five, are used in a simple accord which represents up to 80% of the formula. Around this simple structure are arranged a number of other materials including bases and naturals, which provide the richness and complexity necessary to complete the identity of the perfume. Perfumes created in this way may lack something of the aesthetic quality of earlier perfumes but have the advantage of remaining more-or-less unchanged in odor from their initial impact until after many hours on the skin.

A typical example of such a perfume is Trésor, in which 80% of the formula is made up of four ingredients: methyl ionone, Iso E super, Hedione, and Galaxolide, all materials that seem to lend themselves particularly well to this type of formulation. Frequently in such perfumes the traditional fresh top note is entirely missing, being replaced by trace amounts of intensely powerful materials. This type of structure allows the essential character of the fragrance to come through immediately on application to the skin, rather than having to wait for the top notes to begin to evaporate. This is an important requirement in a perfume that has to be selected from a number of others in a busy department store or airport.

Examining the structure of a perfume in more detail, we may conveniently consider the main components under three separate headings: the perfumery accord; the relation between top, middle, and base notes; and the balance between simplicity and complexity.

THE PERFUMERY ACCORD

One of the most useful exercises that student perfumers can do is to take two materials, beginning with those of similar volatility, and blend them together in every proportion from 9:1, 8:2, 7:3, to 1:9 selecting the mixture that seems to them to be the most aesthetically pleasing—usually the one in which the two materials are smelled in equal intensity, in which neither one predominates. Sometimes the experiment will produce nothing very interesting, but occasionally something remarkable occurs: The mixture takes on a character of its own which is distinct from its component parts. Such "accords," for example, as between patchouli and hydroxycitronellal, or between eugenol and benzyl salicylate, are used over and over again by perfumers and have the remarkable ability to retain their character within complicated blends of other materials.

From here students may go on to make accords between three or four materials, or between materials of different volatility, guided by their own intuitive sense, so as to arrive at an aesthetically pleasing balance. For example, ylang may be added to the combination of eugenol and benzyl salicylate to produce the basic note of carnation that occurs in L'Air du Temps.

In his articles entitled "A Method of Creation in Perfumery" Carles (1961) recommended the systematic study of important materials such as oakmoss, patchouli, and methyl ionone by making combinations of them with a great variety of products, both defined synthetic materials and naturals as well as bases and specialities, beginning with two or three materials and then going on to five or six. Much of the work was purely empirical, done by combining the materials in a number of mathematical proportions, for example, 1:1:1, 3:1:1, or 3:3:1, before allowing the inspiration and originality of the student to select and modify the final accord. The results of these experiments needed to be learned and memorized. Carles himself claimed to have carried out over a thousand such experiments with oakmoss alone. However, useful though this method can be as a learning exercise and as a way of finding new and often surprising relationships between materials, it is a mistake, when creating a finished perfume, to be bound too rigidly by such proportions, since by adding further materials to the accord we may need to modify the original balance. When we reach the full complexity of a perfume, we need increasingly to be guided by intuition and further experiment to arrive at the most aesthetically pleasing balance within the formula. It was the great perfumer Arturo Jordi-Pey, a man of enormous experience and culture, who used to describe perfumery as "the art of the intuitive accord."

We must admit that our repeated use of the expression "aesthetically pleasing" is something of a descriptive cop-out, for we do not know exactly what causes this synergism among olfactory materials. Nor is it easy to describe, but materials working together in a finely balanced accord seem to produce an olfactory resonance as harmonically satisfying to our sense of smell as the sound of a perfectly balanced orchestra is to our ears.

That this olfactory synergism exists is equally well-known in the world of taste. The combinations between certain wines and cheeses, or between saffron and seafood, are a constant source of pleasure and surprise. Incidentally, most perfumers are also connoisseurs of fine food, and many are excellent cooks.

Thought of in this way a perfume may be seen as an accord between all its ingredients, which come together to produce its unique identity. One of the most exciting moments for a perfumer is when the composition on which he or she is working begins to take on such a personality, which can then be developed and refined into a finished perfume. To reach this stage, the perfumer will frequently bring together a number of lesser accords as building blocks in the final composition. At the heart of most successful perfumes is what we may think of as the main structural accord which defines the essential character of the fragrance, whether it be a chypre or an oriental, or a modern perfume based on an accord of four materials. To this framework are added modifying materials and other accords, often in the form of preblended bases. All are fitted together to form the closely knit structure that represents the final perfume. Sophia Grojsman has aptly described the process as being like the coming together of the squares of a Rubik's cube.

THE RELATION BETWEEN TOP, MIDDLE, AND BASE NOTES

Perfumery materials differ widely in their volatility, from those that last only a few minutes on a smelling strip to those that are still there after several weeks. It is usual therefore (Poucher 1955; Carles 1961) to divide materials into three groups: the base notes which are the most long lasting; the middle notes, or modifiers, which are of medium volatility; and the top notes which are the most volatile. The balance between these three groups of materials in a formula is of great importance to the way in which the perfume diffuses during its evaporation and to its aesthetic quality.

Diagrammatically Carles represented the structure of a perfume in the form of a triangle divided horizontally into three sections representing the top, middle, and base notes, with the proportions between them as shown in Figure 9.1. A perfect example of a perfume formulated in this way is L'Air du Temps.

The division into top, middle, and base is of course somewhat arbitrary depending on where we draw the line between the three types of product. There is also considerable difference of opinion between perfumers as to the interpretation of evaporation tests carried out on smelling strips. Helmut Führer (1970), for example, places benzyl salicylate and methyl ionone among the top notes, while Carles (rightly in our opinion) includes them among the base notes. The curious placing of benzyl salicylate by Führer may perhaps be explained by its apparent weakness of odor particularly after any trace impurities have evaporated. Methyl ionone we would place near the top of the base

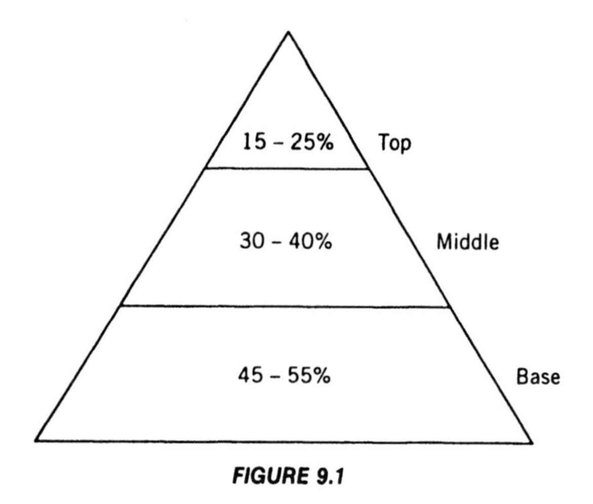

FIGURE 9.1

notes working in many perfumes as a bridge between these and the middle notes. Many natural products are also difficult to place because of the varying volatility of their constituents. It is usual with these to consider only the tenacity of the most characteristic note of the product. This again is open to a wide range of interpretation. Ylang oil, for example, contains constituents that work throughout the evaporation of a perfume. Generally speaking, most of the traditional fresh notes such as the citrus oils, linalool, linalyl acetate, and lavender are placed among the top notes, together with more modern materials such as dihydromyrcenol, rose oxide, and *cis*-3-hexanol. Middle notes include the majority of floral notes such as terpineol, the rose alcohols, and many of the most important chemicals used in the compounding of muguet and jasmin. Eugenol, the essential ingredient of carnation as well as of many spicy oriental notes, comes somewhere near the bottom of the middle notes, as do amyl and hexyl cinnamic aldehyde. Often it is more convenient to think of these materials as base notes. Among the undisputed base notes are oakmoss, patchouli, most of the woody materials, the musks, and vanillin.

In a well-constructed fine fragrance the heart of the perfume needs to be contained within the lower part of the composition, since this part will remain on the skin for many hours after application. Many of these long-lasting materials, as pointed out by Carles, are rather unpleasant when smelled initially and it is part of the function of the modifiers and finally the top notes to subdue and round off the character of these products in the final accord.

It is useful to note that the retention times of materials determined by gas chromatography, using a carbowax-type polar column, give a fair indication of their relative volatilities within a compound. However, most perfumers develop their own way of thinking about volatility based on their personal experience and method of work. Too much of a scientific approach can be misleading, and the student perfumer will, in the words of Carles, "soon attain unexpected proficiency by forgetting any technical information he may have, and by establishing 'his' classification for himself."

The relation between top, middle, and base notes is one that can be applied to all perfume formulation, though the actual proportions given in the triangle of Figure 9.1 would apply to a perfume to be diluted in alcohol at between 12% and 18%. Owing to the effect of the alcohol itself, the optimum dosage for a perfume compound is not necessarily the same for every compound, and this needs to be established by trial and error. In men's fragrances, which are usually diluted at a lower level, there is frequently a much higher proportion of the more volatile materials, necessary to give the freshness that is required in such products. As we have already noted, in many modern perfumes the proportion of the top note has been greatly diminished by the use of much smaller amounts of intensely strong materials. In functional products such as soaps and cosmetic creams the product base can equally play an important role in influencing the most appropriate balance between the components of the perfume. The special problems relating to such products will be discussed more fully in Chapter 13. As always in perfumery, the perfumer must ultimately be guided by experience and aesthetic judgment. The "rules" are there as a guide rather than to impose a fixed method of working.

We have already noted in passing that certain types of odor tend to be represented more heavily in one of the three groups than in the others. This can be explained by the relationship, on the one hand, between molecular structure and volatility and, on the other, between molecular structure and odor. Acetates, for example, are generally more volatile than their equivalent alcohols and also fresher in character. Most typically fresh and green notes fall within the upper part of the spectrum, while woody, sweet, and animalic notes tend to fall within the lower half. One of the challenges to the perfumer is to achieve a harmony between the different levels of the perfume in which the character of the perfume is carried through from one part to another. This is particularly difficult in the type of men's product that contains a high proportion of fresh and comparatively volatile materials. Eau Sauvage is a wonderful example of the way in which such a problem has been solved, with the fragrance maintaining its essential character throughout its evaporation.

THE BALANCE BETWEEN SIMPLICITY AND COMPLEXITY

As we saw in an earlier chapter, many of the most beautiful fragrances that occur in nature, for example, those of flowers, are made up of a combination of hundreds of individual ingredients. Of these comparatively few make up the main structure of the oil, while all the other ingredients make up the small remaining part. This balance between simplicity and complexity appears also to play an essential part in the structure of a well-made perfume that has both identity and that indefinable something we think of as quality. Again we do not need to understand why this should be (although it may be interesting to speculate); it is something that we know from experience and that has become part of well-established perfumery technique.

The English perfumer and writer W. A. Poucher used to tell his students that there were only three rules in perfumery: simplicity, simplicity, and again simplicity. Of course he was working at a time when it was not unusual to use a large proportion of flower products in the composition of a perfume, and it was these, each containing hundreds of ingredients, that brought the necessary complexity to the composition. Poucher was also a fine landscape photographer and applied the principle of simplicity equally to the composition of his pictures, allowing the detail and texture to provide the quality that was the hallmark of his work.

Another of the most valuable exercises that a student can do, and one for which we are grateful to our colleague Martin Gras, is to recreate the essential character of a perfume from no more than 10 or 12 materials. This teaches not only the underlying accords that go to make up the perfume but also shows what can be achieved with comparatively few materials. This technique is also invaluable when trying to simplify a new formula or trial match that has become overcomplicated and unworkable—to go back to the essential structure and to start again from there.

Although a well-made perfume needs to be constructed around a strong and comparatively simple accord, it is the presence of a multitude of other materials that completes the full character of the fragrance, giving it the roundness and aesthetic quality. Of course we are not advocating complexity for its own sake, and the perfumer who continues to add more and more materials in the hope that they will somehow cover up the imperfections of his or her original accord is more than likely to be disappointed. The complexity of a perfume must be related to its overall structure with no material added just for the sake of complexity. Each one must be there for a purpose, making a contribution to the final accord that makes up the finished perfume. The perfumer should know how each material in the formula works in relation to a creative idea—or leave it out.

There is a marvelous, and authentic, moment in the film *Amadeus* when the Emperor, struggling to make some sort of informed comment at the end of one of Mozart's operas, is prompted by the composer's rivals to complain that there had been "too many notes!" Mozart, with the sublime assurance of the truly gifted, replies in astonishment that there were just as many notes as were needed: not too many and not too few. The ability to judge the appropriate balance between the decorative complexity of a composition, whether it be in music, painting, or perfumery, and its underlying structure is one of the marks of true creative greatness.

FIGURE 9.2

This relation between simplicity and complexity in perfumery may be represented by the diagram in Figure 9.2 which shows that each material not only plays its part in forming the overall character of the perfume (at the outside of the circle) but also contributes to its internal structure and harmony.

Complexity in a perfume can often best be achieved by the use of bases, or subcompounds, and by the use of natural products. The different types of base and their use are discussed more fully in the following chapter, but they may be thought of in the present context in the same way as essential oils. A good base is one that introduces into a composition a particular character that makes a contribution to the overall accord. At the same time it makes a contribution to the richness and complexity of the fragrance, often by introducing trace materials, which because of their olfactory strength cannot conveniently be added to the formula individually. It is not unusual for certain types of base, such as green, fruity, and animalic accords to be made up in solution before adding them to the composition. This has the additional advantage of making the perfume virtually uncopyable.

However, a perfumer needs to justify the complexity of a formula and the use of bases against the demands of his or her company in keeping compounding costs to a minimum. To include in a formula half a dozen or more bases, all of which contain many of the same materials is quite unjustified from the point of view of production. The perfumer should try, wherever possible, to use existing bases rather than trying to work only with bases of his or her own making. Much will depend on the importance and likely offtake of the project. A company is much more likely to be tolerant of the use of specially created bases in a major perfumery success than in a compound for which the demand is less than 100 kg a year.

Some companies, in order to keep manufacturing costs to a minimum, adopt a policy of rewriting, or "exploding," the perfumer's formula into a single linear form to get rid of the bases. A cutoff level is then applied removing any material of below a certain percentage. This is a procedure that needs to be used with discretion, depending upon the type of product for which the perfume is destined, and the final formula should always be under the control of the perfumer.

Much of what we think of as quality in perfumes is associated with the complexity that comes from the use of natural products, and in particular with those derived from flowers. Jean Carles used to describe jasmin absolute as being to perfumery what butter is to haute cuisine: The effect of margarine is never quite the same. We may formulate inexpensive fragrances which are instantly recognizable as being, for example, a Chanel 5 or L'Air du Temps type, but without the addition of rose and jasmin absolutes the perfumes lack the quality and aesthetic appeal of the originals. Of course few perfumers today can afford to use such materials in the quantities found in the great perfumes of the past. But naturals remain an indispensable part of those perfumes that we associate with beauty of fragrance and quality. Even in the formulation of fragrances for use in functional products, the use of small amounts of natural materials, such as lavandin and geranium, can have a remarkable effect on the performance.

The complexity that comes from natural products plays an important part in the composition of all fine perfumes. Although the underlying structure needs to be simple, it is the complexity that gives a perfume that aesthetic quality admired by the discerning consumer. However, to end this chapter on a rather sad note, the average consumer of today is frequently more concerned with performance, in terms of strength and staying power, than with the beauty and subtlety of a perfume.

10

The Use of Bases

Bases may be thought of as the prefabricated building blocks of perfumery. They may be as simple as an accord between three or four materials or nearly as complicated as a complete perfume. A base should have a well-defined character, since it is an essential structural element of the perfume's composition.

Perfumers vary widely in the extent to which they use bases in the creation of their perfumes. Some regard the making of bases as one of the most important aspects of their creative work. They embody in their bases their most original ideas, building them into the otherwise classical structure of their perfume to provide much of its essential character. Michel Hy used to tell his students that it could take him a year to make a good base, but once complete he could use it to make a perfume in a week. And sometimes he would do it in less.

Other perfumers prefer to work on perfumes with "open" formulas, using bases only to give special effects such as a green or fruity note. Many modern perfumes would appear to be constructed in this way with most of the formula made up of simple materials.

Many of the earliest bases, some of which are still widely used, date from the early part of this century, when raw material supply houses were less involved than they are today in the formulation of finished fragrances. Often the new synthetic materials produced by such companies would first become available to perfumers, working independently or employed by the fashion houses, wrapped up in the form of speciality bases. In this way the exclusivity of such materials was preserved, while providing perfumers with a whole new palette of olfactory notes with which to work. Specialities such as Parmantheme based on nonadienal, Florizia based on allyl ionone, and Mousse de Saxe based on isobutyl quinolene all date from this period. This practice still continues today, not so much to hide the identity of the captive material (which can usually be revealed by analysis) but as a means of selling the material in a way that can be more readily incorporated into a formula. Among more recent introductions are Vertralis based on Vertral, Cassis 281 based on Buccoxime, and Dorinia based on beta damascone.

The availability of new synthetic materials during the first decades of the century also provided the inspiration for the creation of bases that attempted to duplicate the fragrance of flowers, many of which were already used in perfumery. One of the first to be imitated in this way was lilac, based on a combination of terpineol, heliotropin, and cinnamic alcohol. Benzyl acetate, amyl cinnamic aldehyde, and indol were used as the basis of jasmin; the ionones made possible the re-creation of violet; hydroxycitronellal formed the starting point for muguet; while eugenol, cinnamic alcohol, vanillin, and benzyl salicylate formed the basis for the duplication of carnation. The use of such floral bases and their more modern descendants still forms an indispensable part of the perfumer's technique in the building of perfume creations. Compounding notes for many types of floral base have already been given in Chapter 6.

More recently the great advance that has been made in chemical analysis and the synthesis of complex organic molecules has made possible a much closer approximation to the actual composition and odor of natural flower products. Even so, few of these reconstructed naturals, invaluable though they are, can provide a full replacement for the genuine product. They may best be thought of as specialized bases resulting from a collaboration between the chemist and creative perfumer.

A technique occasionally used in the production of floral bases is to process the compound over the natural product. For example, a neroli compound may be added to the flowers before distillation or a jasmin may be distilled over the spent waxes after making the absolute from the floral concrete. Such *surfleurs* products have a naturalness that cannot be obtained by the compound alone.

Many of the important perfumery accords, first discovered during the first half of the century, were also embodied in some of the famous bases created at that time. A number of these are still available and widely used. The Ambreine and Mellis bases, which we will be discussing in connection with the oriental perfumes, are examples of such accords, as are many of the animalic bases built around the combination of paracresyl derivatives, phenylacetates, and cedarwood.

Some of the most successful bases contain quite simple accords of only two or three materials, perhaps dressed up with a number of auxiliary products. The combination of phenoxyethyl isobutyrate and dimethyl benzyl carbinyl acetate, used in many fruity notes, produces an unmistakable character that survives even when used in trace amounts in a finished perfume. Similarly an ambergris base, made from a simple combination of labdanum, olibanum, and vanilla, brings an unmistakable effect. Such bases, which the perfumer may create for him- or herself, are not only valuable building blocks in the creation of a perfume but are a convenient way of introducing trace amounts of materials that in combination make a special contribution to the character of the final composition.

Another technique much used in the past, though less popular today, was to take well-known products such as methyl ionone, vetiveryl acetate, or hydroxycitronellal—materials that themselves could often form 10–20% of a finished perfume—and decorate them with a number of other synthetic and natural materials to form a base with much of the complexity and "roundness" of a finished perfume. Many such bases, for example, Althenol and Selvone (used in Ma Griffe), were created by Carles to be used not only by himself but also by client perfumers who lacked perhaps the same level of technical proficiency.

One of the most valuable uses of bases today is as a means of introducing into a perfume intensely powerful materials that in isolation can give an unacceptably harsh effect. Many green notes fall into this category. But, by combining them in a carefully worked accord, they can in fact bring a naturalness to a perfume, and one that is also very difficult to duplicate. Fruit bases are another example of this technique. Combinations of powerful fruity materials, often using products such as phenoxyethyl isobutyrate or hexyl cinnamic aldehyde as a carrier, combine to give an effect unobtainable by formulating with single materials. Trace amounts of fruit bases are some of the most widely used in perfumery today.

Yet another type of base is that in which the essential character of an existing perfume is recreated without developing the full complexity of the original. This can then be used to give a twist to the character of a new creation or as the starting point for a "within the area" match. Such bases, even if written out into the final formula, can be of special value in the creation of hybrid fragrances, such as between Anais Anais and Paris, or between Calandre and Rive Gauche, which are frequently used for deo-colognes and other toiletry products.

A similar technique that can be used by a perfumer working with GC information is to extract a number of identified materials that make up one aspect of a perfume—for example, all the rose notes—and bring them together in the form of a base. Occasionally the perfumer may in fact be recreating the actual base used in the original.

Some perfumers, less commendably than those who create genuine bases with which to work, will extract a "heart" from their finished composition, in the form of an artificial base for the purpose of secrecy, concealing important aspects of the composition from compounders, or from other perfumers who may get a glimpse of the formula. This practice, especially when several such pseudobases are involved, can cause enormous problems when it comes to production or to the modification of the perfume for other applications. Some companies legitimately use this system of hearts when transfering formulas to branch companies, but it is a practice that should not be encouraged in perfumers.

Perfumery style today, demanding a greater initial impact based on a simpler type of formulation, has reduced the use of the type of base that represents a high proportion of the finished composition. But floral bases and those that bring a strong positive character to a perfume are still widely used, and a knowledge of them is an important part of a perfumer's training.

Part III

Studies in Fine Fragrance

11

The Descent of Perfumes

There is an old English proverb that says that a bride should include in her wedding attire "something old, something new, something borrowed, something blue." If we use the word "blue" in its somewhat erotic sense, then this is not a bad description of the attributes of most successful new perfumes. Few perfumes that come on the market today are entirely original, and if they were, it is unlikely that they would be accepted by the consumer. They are derived from earlier perfumes that have a proven record of market success, and the successful launch of a major perfume is likely, within a very short time, to be the inspiration behind numerous other fragrances that are more or less closely related to it. Some may be equally prestigious, while many will be adaptations for a lower-price area of the market. The same process can be seen at work in the world of fashion. What is new in Paris one year is likely to find its way into the chain stores the next.

This is not to suggest that perfumers are any less creative than artists working in other disciplines. The history of all forms of art can be thought of in terms of evolution: the derivation of what is new from what has gone before, a process of gradual change interspersed by the occasional great leap forward of imaginative inspiration. Without Haydn the symphonies of Beethoven would not have been written, nor without these the symphonies of Brahms. Without the paintings of Cézanne there would have been no Picasso—or not the one with whom we are familiar. All artists are influenced by the period in which they work, drawing upon the past while responding to the cultural changes that surround them. We may see, in retrospect, a continuity of influence stretching back over many generations of creative work. Similarly in perfumery we can trace the origins of many of today's formulas back to the perfumes created at the beginning of the century. The driving force behind this evolution can be seen to have been the interaction between the availability of new materials, the genius of the creative perfumer, and the changing demands of the marketplace.

There is also a parallel to be drawn between the evolution that has taken place in perfumery with the evolution of living things by the process of variation and natural selection. The emergence of modern perfumery may be said to have been made possible at around the turn of the century by the work of the organic chemist. There were new materials to work with, new ideas to be explored, and a marketplace largely unconditioned by anything that had gone before. Into this environment, as at the dawn of natural evolution (in the so-called Cambrian explosion), there appeared a great diversity of types. Some were perfumes of startling inventiveness, aimed at a highly specialized market dominated by the fashion-conscious elite of society. Many of these perfumes disappeared without trace, or have survived as living reminders of a bygone age, but from among them emerged certain combinations of materials (the genetic material of perfumery) that proved more successful in competing for a market share than others. Not only were they successful in themselves, but they were also capable of being adapted, by the introduction of new materials and ideas, to the changing conditions of the market environment. In the natural world it was the success of their underlying structures that gave rise to such groups as the molluscs, insects, and vertebrates. In perfumes it was to be the chypre, oriental, and floral-aldehydic accords, among others. Through a combination of creative variation and consumer selection, they came to dominate the market, producing whole families of perfumes that could be adapted to a number of different product types and radiate out into every niche in the fragrance market.

Because of this evolutionary pattern in the development of perfumery, it has become usual within recent years to classify perfumes in the form of a genealogy, or family tree. Such genealogies are designed to show the relatedness of perfumes based on their underlying accords. We may trace, for example, the descent of many modern floral-oriental perfumes from L'Heure Bleue and L'Origan, through Oscar de la Renta and Vanderbilt, to such perfumes as Poison and Loulou. All these are based on a characteristic accord between ylang and eugenol (carnation), Schiff bases (orange blossom or tuberose), methyl ionone (with the exception of Poison), and vanillin, with heliotropin and coumarin. The actual relatedness of perfumes can vary from one of being a "me-too" copy, through one of clear descent (the relationship of Fidji to L'Air du Temps), or to a largely original perfume that falls within an existing family (Ysatis as a modern interpretation of the chypre accord). Occasionally it is possible to see the influence of an existing fragrance on a new perfume in a different family. In Obsession, a modern oriental perfume descended from Shalimar, the top note has clearly been borrowed from Alliage, a fruity-green chypre perfume. Totally new perfumes, apparently unrelated to what has gone before are rare indeed. Eau Sauvage, although stemming from the classic Eau de Cologne accord, can be seen in retrospect as having been the starting point for a new family of "fresh" perfumes; one of the great imaginative leaps forward. Similarly Coriandre with its emphasis on patchouli, with hedione and rose as the dominant floral notes, can be thought of as the beginning of a new family within the chypre area, leading on to such perfumes as Aramis 900, Paloma Picasso, and Knowing. The composition of many of these perfumes will be discussed in greater detail in the following chapter.

It is not unusual to find the introduction of a new perfume that relates back to one from a much earlier period, as in the relationship between Oscar de la Renta and L'Origan. Perfumery fashion tends to go in cycles. Families may remain dormant for a considerable time before being reintroduced, often by the creation of a perfume inspired directly by a model from the past but using many of the new materials that have become available within the years that separate the two.

Many of the greatest masterpieces of perfumery draw upon the construction of earlier perfumes. Even the great Jean Carles based much of his work on the already well-established chypre and fougere accords, introducing into them a degree of originality rarely found in modern perfumery. Few perfumes today can match the astounding use of materials found in Canoe or Ma Griffe. The study of such creations, seen in the context of their place in the evolution of perfumery families is one of the most worthwhile exercises that a young perfumer can undertake.

Modern methods of analysis have not only made it possible for students to acquire a greater knowledge of the perfumes of the past but have also brought about a much greater emphasis on the creation of clearly derivative perfumes, both those that genuinely bring something new and those that are little more than me-too copies. The combination of market forces and GC analysis has resulted in an explosive acceleration of the rate at which the perfumery evolution is taking place.

This is not to deny that many great and largely original perfumes have been produced over the past 20 or 30 years. Among these must be included many of the perfumes of Estee Lauder, such as White Linen and Alliage, as well as such perfumes as Diorella, Eternity, and Samsara. Such perfumes are often difficult to classify or relate directly to any one perfume of the past. Sometimes such a perfume will represent a one-off, falling outside the mainstream of perfumery descent. Occasionally however, it will be seen in retrospect to have been the beginning of a genuinely new direction, starting a whole new family line. Among recent perfumes, Trésor, representing a new style in perfumery technique, can already be seen as the inspiration behind a number of subsequent introductions including Escape, Dune, and Volupté.

Although the genealogical approach to the classification of perfumes is widely accepted by perfumers as a valid way in which to interpret some of the creative influences under which they work, it is a mistake to be too categorical in trying to draw up what amounts to a family tree of supposed relationships. Some perfumes, although we cannot think of them as among the great originals, are almost impossible to place in such a classification. Ombre Rose may be thought of as related to Chanel No. 5, as a floral-oriental, or even as being close to, and perhaps inspired by, Johnson and Johnson's famous Baby Powder.

Apart from the fine fragrances that have been mentioned so far in this chapter, there exists an enormous market for perfumes designed for toiletry products such as deo-colognes. Many of these are more-or-less copies of successful fine fragrances adapted for a particular market and price. Very often, however, they can be seen to be hybrids between two closely related perfumes, for example, between L'Air du Temps and Fidji or between Anais Anais and Paris. Such fragrances, although not very original, fit the market profile in having an assured acceptance with just sufficient individuality to give it product identity. Trying to create hybrids between perfumes belonging to unrelated families is rarely successful. But, as we have seen in the relationship between Obsession and Alliage, it is sometime possible to borrow some of the auxillary notes, such as a special green or fruity accord.

A difficulty that the student perfumer faces when examining some of the older perfumes is knowing the degree to which many of these have been altered since their original introduction. Sadly, some of the great perfumes of the past are no longer what they used to be. The cost and availability of raw materials may have necessitated adjustments to the formula. The changing quality of raw materials has been one of the most insidious ways in which many of the great perfumes have degenerated into shadows of their former glory.

Sometimes a perfume will be relaunched using a deliberately reformulated compound designed to adjust to market prices, and to bring the fragrance more into line with current requirements without sacrificing too much of the original character or quality. Others have been marginally changed over the years with the introduction of small amounts of new synthetic products, such as jasmin, sandalwood, and musk materials, so as to reinforce the character of the original and to give some of the strength demanded by the current market—but, no doubt, with some eye to a bit of incidental cost cutting.

Occasionally the fragrance of an old perfume will be radically changed, possibly because of problems with the availability of raw materials or sometimes simply because sales of the original perfume have declined. Je Reviens, a unique survivor of an earlier style of perfumery, and one of the best loved among its committed users, has sadly undergone such a change. It is now unlikely to please either its established customers or to win any new converts. A great loss to perfumery.

Besides the genealogical approach to the classification of perfumes, which is particularly relevant to the theme of this chapter, a number of other classifications exist based on different criteria and with other areas of usefulness. The *Classification des parfums,* published by the Société Technique des Parfumeurs de France (STPF) in 1984, deliberately avoids the genealogical approach, preferring simply to group together perfumes that fall within the same generalized area as defined by their odor description. This approach, which is also followed by Haldimann and Schuenemann in the *Dragoco Hexagon of Perfumes* (1988), is widely used at a marketing level within the industry to demonstrate perfumery trends or, as in the case of the STPF, simply as a catalogue. Frequently the results of such a classification follow closely those of the genealogical approach. One area of difference often lies in the placing of many of the so-called green perfumes. Because of their very individual character such perfumes are frequently grouped together under the title "green-floral." In the STPF classification this brings together Fidji, a direct descendant of L'Air du Temps, Alliage, which is at least arguably a floral-chypre, and Chanel 19, a perfume belonging to its own genealogical family of intensely woody floral-green perfumes. These three perfumes have little in common except their green character, and even this is based on different combinations of materials. Beautiful, on the other hand, while reputedly derived from Chanel 19, fits much more readily into the descriptive type of classification, since it has little of the green character that lies at the heart of its forebear, and rather more in common with powdery floral notes such as Eternity.

Another approach, adopted by J. S. Jellinek (1992) in his *Map of Perfumes,* is to classify perfumes on the basis of consumer perception. Consumer perception in many cases differs considerably from that of either the perfumer or fragrance specialist within the industry. Such a classification is valuable in the positioning of new perfumes being launched onto the market.

In the following chapter we will consider the compositions of some great perfumes of the past and present, looking in particular at the relationship between perfumes belonging to the same generally recognized families. Such comparisons, whether or not they reflect the actual derivation of the perfumes involved, are invaluable to the student perfumer in understanding the technical basis of creation and formulation. In other cases, where a perfume does not belong to any clearly recognized line of descent, we will adopt the purely descriptive approach to classification.

12

Selected Great Perfumes

Not so many years ago it would have been impossible to discuss, in writing, the detailed composition of the great perfumes. The actual formulas were, and still are, in the hands of their creators or the companies for which they work. Although such formulas have occasionally found their way into other hands, the information they contain continues, quite rightly, to be regarded as confidential.

One could of course learn a great deal about their composition through the painstaking effort of making a more-or-less close olfactory match. But one could never be quite sure how faithfully the formula of the match actually reproduced the original. Moreover the match itself represented such a large investment in time and effort that no one would think of publishing the information it contained.

Today, however, with the ready availability of modern methods of analysis, extensive information has become the common knowledge of perfumers. A good GC-MS printout can provide not only the main structure of a fragrance but also detailed insights as to most of its constituents, both natural and synthetic. Using this as a starting point, an experienced perfumer can go a long way toward reconstructing the formula as originally conceived by its creator.

One can never be sure, of course, as to what went on in the mind of the creator. Such is the nature of art that even if one were to ask him or her, one might not get a very coherent reply. The creator is often quite unaware of the experiences that influence his or her work. We might, on analyzing a perfume, find it to be related structurally to

Madame Rochas or to Calandre. If we were to point this out to the perfumer who made it, his or her response might quite genuinely be one of surprised acknowledgment.

Different perfumers have different ways of looking at perfumes, and no doubt some of our readers will disagree with many of the ideas set out in this book. Perhaps the creators of some of the perfumes discussed will be among them. We would accept such disagreement with equanimity, for it is not our purpose to promulgate the idea of a one and only true family tree of perfumes. Rather, we want to use these perfumes, which can be smelled and worked on by students, as a starting point for a discussion of the use of materials and to provide insights into the techniques and creative methods used by the great perfumers, past and present.

The information given on the following pages is based on data coming from GC analysis combined with further work carried out in conjunction with our students. Most perfumes exist in more than one form, being sold not only as a concentrated extrait, usually referred to as the "perfume," but also in more dilute versions such as "toilet water" and "eau de parfum." Such products are frequently based on rather different formulas from those of the original perfume; more modern and less expensive ingredients are used to recreate the same idea, but in a way that is more commercial and more suitable to the lower concentration. With this variation between the formulas of different line products, and the changes that have taken place in the formulas of many older perfumes, it is not always possible to talk of the one-and-only formula for a particular fragrance. Although most of the information on which we have based our studies comes from the analysis of the extrait versions as they exist today, in the older perfumes some of the ideas are based on what we believe to have been the intentions of the original perfumer. In other cases we have also used information derived from the analysis of the more dilute versions.

It has not been our intention to provide complete reconstructions, or indeed formulas, for the perfumes under discussion. To do so would be of little value to the serious student wishing to develop an ability as a perfumer, or to learn from the work of matching. Nor would we wish that such formulas be used for the unethical marketing of copycat, or at worst, counterfeit products. It remains for the individual perfumer or student, who wishes to do so, to carry out the work of reconstruction and to arrive at his or her own interpretative formulas for the great masterpieces of perfumery. Where percentages are given, these should be regarded only as guidelines to the main structure of the perfume and as a starting point for further work.

In the selection of perfumes to be discussed, we have concentrated on the perfumes that are at the origin of the major families as well as other outstanding representatives of those families. We have chosen to restrict our discussion largely to feminine perfumes.

THE FLORAL SALICYLATE PERFUMES: L'AIR DU TEMPS, FIDJI, ANAIS ANAIS, PARIS

In this group we will be looking at four perfumes, predominantly floral in character, based on salicylates in combination with woody notes and musk. This underlying accord has proved to be one of the most successful in perfumery, providing the starting point for many varied and wonderful creations.

L'Air du Temps of Nina Ricci must be regarded as one of the most important perfumes ever made. Not only has it enjoyed a great and enduring commercial success, but its influence can be seen behind the creation of a number of subsequent perfumes, such as Fidji, Charlie, Chloe, Anais Anais, and Paris, which can be thought of as forming a more-or-less well-defined family. It has also been the inspiration behind innumerable trickle-down fragrances, particularly for toiletry products and cosmetics.

L'Air du Temps

Created in 1948, L'Air du Temps is a perfect example of a perfume belonging to what we have described as the middle period of perfumery, formulated around a structure composed of well-defined top, middle, and base notes. Its extraordinary simplicity, relying on natural products to give complexity and richness, not only makes it one of the most distinctive of perfumes but also a natural starting point for many derivative and often much more complex types of formula.

During the early years of the century a number of successful perfumes had been made based on salicylates, usually on a combination of amyl and benzyl salicylate. Among these were floral-aldehydic perfumes such as Quelques Fleurs and Fleurs de Rocailles with a dominant lilac note, and Je Reviens based on narcisse and jonquille. Many of these perfumes also contained a combination of eugenol and isoeugenol as part of a carnation complex, and it was these in combination with benzyl salicylate which were to form the inspiration for the creation of L'Air du Temps.

Salicylates, as part of the "mellis" accord (discussed later in this chapter), had also been used in perfumes such as Blue Grass and

Moment Supreme, in combination with jasmin, rose, eugenol, clove, coumarin, and vetiveryl acetate, and with oakmoss in fougeres such as Fougere Royal, and Canoe. Few of these perfumes have survived in their original form, and amyl salicylate has largely fallen out of favor as a major structural component in fine perfumery, its place having been taken by the more elegant *cis*-3-hexenyl salicylate.

At the heart of L'Air du Temps lies the classic accord, 4½ : 1, between benzyl salicylate (15%) and eugenol. The large proportion of these two materials together with ylang and isoeugenol provide the essential carnation character that dominates the perfume throughout its evaporation.

The base note also contains the important accord between methyl ionone (10%), vetiveryl acetate, sandalwood, musk ketone, and originally musk ambrette. These materials together with the carnation make up the immediately recognizable central character of the perfume.

The middle part of the perfume is a bouquet of floral notes reduced to their simplest components. Terpineol for lilac, styrallyl acetate for gardenia, phenylethyl alcohol for rose, hydroxycitronellal (10%) for muguet, and benzyl acetate and amyl cinnamic aldehyde for jasmin.

The top note (14%) is a classic mixture of bergamot and rosewood, together with their naturally occurring ingredients linalool and linalyl acetate.

Although the essential character of the perfume is built around the ingredients already mentioned, most of its richness and quality comes from the use of jasmin and rose absolutes. By the addition of these two materials many hundreds of individual ingredients are added that envelop and decorate the main structure of the perfume. Here we have a perfect example of classical simplicity combined with the complexity that comes from the use of fine natural materials.

In addition to the products mentioned so far, which form the essential structure of the formula, we may add a number of auxiliary materials that, despite being present in only trace amounts, have a remarkable effect on the performance and aesthetic quality of the fragrance. Aldehyde C11 undecylenic brings additional impact, blending perfectly with the styrallyl acetate and acting as a bridge between the top notes and the rest of the perfume. A further demonstration of the importance of trace materials is seen in the completion of the carnation character by the addition of trace amounts of vanillin, heliotropin, and orris. The use of such materials in this way is an important part of the know-how of perfumers in giving finish and elegance to their creations.

Vanillin is one of the most important, but also one of the most difficult materials to use in perfumery. In an oriental perfume it may frequently form 10% or more of the formula, whereas in another type of perfume as little as 0.5% can totally obscure an otherwise finely balanced accord. But used with discretion, it can have the effect of smoothing out the roughness of a composition, and adding a touch of sweetness, without greatly altering the essential character. When using vanillin in this way, as an auxiliary material rather than as part of the main structural accord, it is advisable more-or-less to complete the work of composition before adding it, carrying out several trials to establish the most effective level. It should never be used just to cover up the faults of a badly put together perfume: Once vanillin has been added, it is often difficult to assess the balance between the other materials.

Vanillin would appear, like many of the animal notes, to act as a trigger to our awareness of smell. Trace amounts in a formula will not only bring out the character of a perfume but raise its level of impact. It appears to work in a perfume in very much the same way as a pinch of salt in cooking. Most dishes, even those that we do not immediately think of as being salty, benefit by its use, but too much can be a disaster. As an ingredient of most carnation bases, as well as of amber notes, it finds its way into nearly every perfume.

Similarly orris, when used even in minute quantities, can have a remarkable effect both on the lift and finish of an already well-composed formula.

Fidji

During the period that elapsed between the creation of L'Air du Temps and Fidji in 1966 there had been an enormous increase in the number of synthetic materials available to the perfumer. New materials frequently provide the inspiration for creative work, and companies, not surprisingly, encourage their perfumers to use such products in their formulas. Even if we had never examined the composition of Fidji before, we could be confident of dating it to within a few years of its creation just by looking at the materials it contains. Although it is not always possible to see any one perfume as being directly derived from another, the link between Fidji and L'Air du Temps is generally accepted by perfumers as being one of direct descent. Throughout the main structure that we saw in L'Air du Temps there has been a deliberate replacement of old materials, either completely or in part, by new. But the balance between the various types of material, allowing for their different comparative strengths, remains almost unchanged. Fidji may be thought of as a L'Air du Temps in which the dominant floral character of carnation has been largely replaced by a "green" hyacinth-jasmin complex. This shift of emphasis is reflected throughout the structure of the perfume.

Among the most important of these structural differences are the introduction of *cis*-3-hexenyl salicylate (6%) replacing part of the benzyl salicylate, and the replacement of hydroxycitronellal by Lyral (8%). Both changes are in keeping with the green, floral direction of the perfume. The woody and musk notes have also been modified and augmented by the introduction of newer materials such as Vertofix (3%) and ambrettolide.

The synthesis of Vertofix (acetyl cedrene) from cedarwood in the early 1960s marked a major step forward in perfumery chemistry, providing perfumers with a fine woody material at a reasonable price, which could be used not only to replace the very much more expensive vetiver derivatives but also had an excellent performance in many functional products such as soaps and shampoos. It is interesting also to find in Fidji a small amount of PTBCHA (Vertenex), another relatively new material at that time, and one more frequently associated with functional products. Here it acts as a link between the woody notes and the fresh green top note.

Ambrettolide, although one of the most expensive of the synthetic musks, when used in trace amounts has a wonderful effect in "rounding off" the character of a perfume, working as much in the top note as in the base.

The main ingredients of the carnation that we found in L'Air du Temps have all been reduced and a corresponding increase made in the jasmin note by the introduction of hexyl cinnamic aldehyde (5%). No doubt if hedione had been available to the creator of Fidji, it would have been used, and Fidji would have been a different, though not necessarily a better, perfume. In keeping with the rest of the perfume, greater freshness has been given to the rose aspect by the introduction of geranyl acetate.

If we examine that part of the composition that corresponds directly to L'Air du Temps, we find the general character to be little changed from the original. What is new are the green and fruity notes, hyacinth, and a slightly mossy character.

Green notes are some of the most important in perfumery, finding their way into almost every type of perfume. But because of their frequent strength they are also some of the most difficult to use. Of the natural products available to the perfumer, galbanum is of outstanding interest because it has wonderful "naturalness" of odor in contrast to many of the synthetic green materials such as phenylacetaldehyde, *cis*-3-hexanol and its esters, Triplal and Vertral, which are intensely powerful and "chemical" in character.

The green note of Fidji is one of exceptional complexity. By using a great many individual materials, in trace amounts, a naturalness can in fact be achieved in which no one "chemical" note predominates. In all probability, these materials are added in the form of several subcompounds or bases, including a hyacinth and a narcisse. Natural narcisse and violet leaf may also be used.

The relation between green and fruity notes is a very important one. Trace amounts of the lower fruity esters such as amyl acetate or ethyl acetoacetate can modify the harshness of many green materials. For the same reason mandarin is widely used in green perfumes, as it is here, particularly in those containing galbanum. Many of the "aromatic" essential oils, such as armoise, Roman camomile, basil, and estragon also have a valuable modifying effect on green notes.

For the mossy character small amounts of oakmoss or foin absolute work well and bring further naturalness to the composition. Other auxiliary products are civet, vanilla, and aldehyde C14.

Many derivative and trickle-down fragrances have been derived from Fidji. Its light, green, floral character together with the tenacity of the salicylates and musk make it particularly suitable for use in toiletry products, especially for shampoos and hair conditioners.

Anais Anais

Although less closely related to L'Air du Temps than Fidji, we may think of Anais Anais (Cacharel 1979) as a perfume created within the same tradition. Again its character is essentially floral, in this case a combination of "white flowers," including jasmin, muguet, lilac, magnolia, tuberose, honeysuckle, and carnation, with a complex of woody notes, musks, and salicylates (6%).

What we notice immediately on examining the composition of the perfume is the almost complete absence of the classical fresh top note that we saw in the earlier perfumes, and the dominance of two materials, hedione (10%) and hexyl cinnamic aldehyde (12%), which with the musks make up some 30% of the formula. This is moving toward the type of structure that we associate with many of the modern linear perfumes. Anais Anais is a fragrance that remains comparatively unchanged throughout its evaporation.

Otherwise, apart from the greater emphasis on floral notes, the general structure remains very similar to that of Fidji. A different combination of musks is used, including cyclopentadecanolide (4%), and cedryl acetate is incorporated into the woody complex. The woody notes are further emphasized by the use of an intensely powerful amber note—a "captive" material that is not generally available. Although the carnation-salicylate accord is again less than in L'Air du Temps, it continues to play an important part in the overall fragrance.

Another important innovation is the introduction of a honey note which includes the use of phenylethyl phenylacetate. This can be backed up by the addition of cire d'abeilles (beeswax) absolute. Although cire d'abeilles varies considerably in quality, it is an excellent material for providing a natural character to an otherwise synthetic floral compound.

The green side of the perfume which is less pronounced than in Fidji, and more floral, is based on phenylacetaldehyde and *cis*-3-hexenyl acetate with perhaps a trace of galbanum. Narcisse absolute may also be used. Other materials that add to the building up of the white flower character are Lilial and cyclamen aldehyde in addition to the hydroxycitronellal and Lyral. The tuberose character can be given by the use of the Schiff base derived from methyl anthranilate and Helional, or by the direct addition of methyl anthranilate. The mossy side of the perfume is represented by Evernyl.

Finally, we should note the use of a relatively large amount of aldehyde C14 (0.3%) which brings to the white flower character of the perfume a touch of exotic fruit.

Anais Anais is another perfume that has been widely adapted for use in many types of functional product including soaps and fabric conditioners. Many of its components such as the salicylates, musks, woody notes, and aldehyde C14 are among the most substantive to cloth, while the floral notes are ideally suited to bring a natural and gentle quality to such products.

Paris

In Paris there is again a close similarity to the structure of L'Air du Temps, but in this case with violet and rose as the dominant florals.

An important innovation has been the use of Iso E Super (6%) as the main woody material, forming an accord with the musk complex and methyl ionone (10%). Methyl ionone is one of the most versatile materials used in perfumery. Although it occurs in a number of different isomeric forms (see Appendix D), the qualities most widely used today are either the gamma (iso-alpha) form or a mixture of isomers in which this predominates. In odor it lies somewhere between floral, woody, and iris, working in a perfume as a bridge between the middle and base notes. It blends with almost every type of material, taking its place as a structural component in many of the great perfume families.

Whereas in L'Air du Temps it is the woody-iris character of the methyl ionone that develops in combination with the vetiveryl acetate, in Paris it forms the basis of a violet accord, with Iso E Super as the woody note. Iso E Super, one of the most important of the newer synthetics, is another chameleon-type product combining woody and amber aspects with some of the character of methyl ionone. Used in Paris as part of the L'Air du Temps structure, it forms a natural link between the other materials and the violet accord.

When creating a perfume, many perfumers like to think in terms of building blocks, working on the violet note, for example, or the green notes, in the form of bases that are then brought together in the final accord. Other perfumers prefer to work with "open" formulas, putting in the individual components that contribute to the various aspects of the perfume's character. (Perfumers who spend too much time working with GC information tend to think exclusively in this way.) Thinking of Paris in terms of building blocks, and we would suspect it to have been created in this way, there is, besides the very beautiful violet accord, a rose complex (10%) reminiscent of the one used in White Linen. The use of DMBCA (2.5%) at an unusually high level in both perfumes is of particular interest. As part of the rose note it works well in conjunction with the underlying woody character.

Hedione, *cis*-3-hexenyl salicylate, and benzyl salicylate (6%) are again important, with Galaxolide (5%) introduced as part of the musk complex in conjunction with musk ketone, cyclopentadecanolide, and Tonalid. It is common practice among perfumers today to use a combination of synthetic musk materials, usually of chemically different types, rather than just one. As with the green notes this complexity gives a more natural and interesting effect.

Paris also contains fruity notes, including alpha damascone and gamma-decalactone, which contrast with the green side of the violet accord. Violet and raspberry, both of which can contain ionones, form a natural association, and a trace of Frambinone can also work well in this context.

Because of the chemical stability of many of the constituents of the violet note, as well as of fruity notes and musks such as Galaxolide, in acid media, fragrances somewhat reminiscent of Paris can be found in household products such as lavatory cleaners. This should in no way detract from our appreciation of Paris itself—a very beautiful perfume.

THE FLORAL ALDEHYDIC PERFUMES: CHANEL NO. 5, ARPÈGE, MADAME ROCHAS, CALANDRE, RIVE GAUCHE, WHITE LINEN

The description "floral aldehydic" is usually applied to the group of perfumes whose origins go back to two of the most successful perfumes created in the 1920s, Chanel No. 5 and Arpège. All contain significant amounts of aliphatic aldehydes in combination with floral, woody, and animalic notes. The earlier perfumes in the group mostly contained a combination of bergamot, linalool, and linalyl acetate in the top note, with ylang leading into the floral notes. These were dominated by rose, muguet, and jasmin, often with lesser amounts of lilac, carnation, or natural tuberose. The base notes included woody materials, such as vetiver and sandalwood, methyl ionone, the nitromusks, and amber. The group may conveniently be divided into two families, depending on whether or not they contain a significant amount of vanillin.

Chanel No. 5 and Arpège

Although the aliphatic aldehydes had been used in a number of earlier perfumes, their dominance (a total of nearly 1%) in Chanel No. 5 marked a milestone in the history of perfumery, and the combination of materials that make up the central accord of the perfume remains one of the most unmistakable, being used also for the perfuming of many types of functional products, from toiletries to room fresheners, where the relative chemical instability of the vanillin (or ethyl vanillin) and the aliphatic aldehydes does not present too much of a problem.

The creation of Chanel No. 5 was based on the bold imaginative use of then available aroma chemicals in combination with fine natural products. Rose and jasmin absolutes were probably used at levels of around 4% or 5%, which would not have been unusual at the time when the perfume was created. In addition there would have been natural musk, ambergris, and civet.

The top note of the perfume contains the traditional combination of bergamot, linalool (rosewood), linalyl acetate, and neroli, with ciste oil as part of the amber note bringing a natural radiance in contrast to the chemical harshness of the aldehydes. A good quality ylang oil is also important in softening the aldehydic character and forms an important 1:1 combination with methyl ionone (8%). The muguet part of the accord is represented simply by hydroxycitronellal (10%), cinnamic alcohol, and styrax. Rose alcohols with a small amount of phenylacetaldehyde add to the floral character. At the base of the fragrance is a combination of woody notes including vetiveryl acetate and sandalwood, isoeugenol, vanillin (1.5%), and coumarin (5%), with a little over 10% of nitromusks.

In Arpège (Lanvin 1927) the floral aspect of the fragrance was further developed by the use of compounded floral bases. Greater emphasis was also placed on the woody notes, including vetiveryl acetate, and on the animalic character, with a corresponding reduction in the vanillin and coumarin.

Although a number of other great perfumes, including L'Interdit (Givenchy 1957) and Topaze (1959), one of the most successful of the early Avon perfumes, can be thought of as being more or less closely related to Chanel No. 5, its very individuality seems to have limited its potential as the begetter of a widely evolved family. In Arpege, however, the influence of vanillin and coumarin within the accord had been greatly reduced, and it was this trend that was to lead to the creation of many of the greatest masterpieces within the group, such as Caleche (Hermes 1961), Madame Rochas, Calandre, Rive Gauche, and White Linen.

Madame Rochas

When comparing Madame Rochas (Rochas 1960) with earlier perfumes such as Chanel No. 5, we are immediately struck by the development that has taken place in the composition of the floral notes as a result of an increased reliance on the use of synthetic materials. A muguet base (15%) using hexyl cinnamic aldehyde and phenylacetaldehyde glyceroacetal has been introduced to augment the hydroxycitronellal (a further 15%), and a special type of rose, here using the absolute made from beeswax, brings added individuality to the perfume. The floral aspect also contains lilac and carnation. Carnation bases, used as auxiliary notes, find their way into a great many perfumes. Containing both eugenol and vanillin, they are a convenient way of adding trace amounts of these materials, both of which have the ability to bring a measure of sweetness and "finish" to a composition. The coumarin is here backed up by heliotropin, rather than vanillin, which complements both the lilac and carnation notes, as well as being an ingredient of certain types of muguet. The combination of ylang and methyl ionone is again present.

As in Arpège the woody note is dominated by vetiveryl acetate (10%), here in combination with a large amount of sandalwood (8%). When using vetiveryl acetate, it was the usual practice to put back some of the unprocessed vetiver oil so as to give greater naturalness to the product, while preserving the fine quality associated with the derived acetate. The woody note is further augmented by the use of cedarwood and by guaial acetate, which, with guaiacwood itself, works especially well in combination with the rose notes. The presence of oakmoss as a major component in the structure of the perfume can be seen as a new direction within the family, influencing such later creations as Calandre and Rive Gauche.

The amber notes used in many of these perfumes are highly complex and frequently difficult to identify. In this case the famous speciality Ambrarome, which was previously used in Arpège, also works well.

Although natural flower absolutes are less important than in Chanel No. 5 and Arpège in establishing the essential character of the fragrance, rose, jasmin, and tuberose may be added, together with tonka, to back up the coumarin (1.5%), and a trace of natural vanilla. A trace also of ambrette seed can give a wonderful smoothness to the composition, working particularly well as part of the rose accord.

Madame Rochas has been the inspiration behind a number of important fragrances used in functional products. The highly successful Camay soap launched in the mid 1950s, although predating Madame Rochas, and often described as an Arpège type, had been built around much the same type of floral, woody, aldehydic complex, also with oakmoss, coumarin, and musks. Vertofix was here introduced as the main woody material with Vertenex (PTBCHA), a comparatively fresh volatile material, reinforcing the methyl ionone, woody, and floral notes. Lavandin was used to reinforce the linalool and linalyl acetate, and the good quality soap base allowed the quantity of aldehydes (similar to those in Chanel No. 5) to play a dominant role. This theme, now generally thought of as a "Madame Rochas" type, continues to be widely used in functional perfumes.

Calandre and Rive Gauche

Calandre was created in 1968 to reflect the metallic theme used at that time by Paco Rabanne in his fashion designs. Although it follows the same general pattern as Madame Rochas, both the rose and muguet notes show interesting developments. The use of rose oxide and diphenyl oxide in combination with geranium, nerol, and geranyl acetate as part of the rose complex provides much of the "metallic" character.

Although the comparatively simple representation of the muguet aspect of the perfume is along the same lines as the one used in Madame Rochas, the greater emphasis on the green notes, using Lyral (4%) as well as hydroxycitronellal, and a significant amount of phenylacetaldehyde, fits also with the metallic theme. The essentially synthetic effect of these green and metallic notes is made acceptable by the use of natural rose oil, one of the most beautiful and irreplaceable products used in fine perfumery.

Although falling into the so-called floral aldehydic family of perfumes, Calandre is one of the least aldehydic, containing a combination of aldehydes C10 and C12 lauric, which together make up about 0.2% of the total formula. Also incorporated into the structure are two important materials, Hedione (6%) and Helional (1%), both of which only became generally available after the creation of Madame Rochas. These two products, previously used together in Eau Sauvage and Diorella, form an important accord. Helional had also been used together with Glycolierral and phenylacetaldehyde glyceroacetal in Printenyl, a speciality base possibly used as such in the creation of Calandre. The accord between hedione and hexyl cinnamic aldehyde (8%) is another that has gone on to play an important role in the creation of many modern perfumes. This combination of synthetic floral notes is rounded off by the use of natural jasmin (3–4%), styrax oil, and additional indol.

The woody part of the perfume is again based on vetiveryl acetate with vetiver (6%) and sandalwood. The oakmoss is backed up by a small amount of Evernyl, a synthetic material created as a result of the study of natural oakmoss, and one that has come to play an increasingly important role in modern perfumery. Coumarin is again present with a trace of vanillin, and cyclopentadecanolide backs up the musk ketone.

Although lacking some of the impact required by today's standards, Calandre remains one of the most beautifully made perfumes. The combination of green, floral, and woody notes has made it the starting point for a number of successful fragrances used in toiletry and cosmetic products.

The comparison between Calandre and Rive Gauche (St. Laurent 1970) is an interesting one. Whereas Calandre had been created for the traditionally elite couturier market in Paris, Rive Gauche deliberately broke new ground in looking toward the younger culture of the Left Bank. Although the structure of the two perfumes is remarkably similar, Rive Gauche has a far greater immediate impact, lacking the subtlety and perfectly rounded character of its forebear.

The floral aspect of the perfume is dominated by a rose complex similar to that used in Calandre but with a greater emphasis being placed on the geranium and aldehydic notes. Cyclamen aldehyde is here used as part of the muguet, with hedione and hexyl cinnamic aldehyde again making up some 15% of the formula. Much of the impact of the perfume comes from the woody character which is here augmented by the use of cedarwood (2%) and Kephalis (3%), a woody amber material of great tenacity, and by a number of other materials in trace amounts. Coumarin and musk ketone are again present.

White Linen

Although Rive Gauche may be thought of as having brought to an end the fashion for classically formulated floral aldehydic perfumes, White Linen (Lauder 1978) further revolutionized the genre with a structure that looks forward to many of today's monolithic perfumes. Although having links to Madame Rochas and Calandre, it follows also the intensely aldehydic character of Chanel No. 5, with nearly 1% of aldehydes but without the modifying influence of bergamot and ylang, or the sweetness of coumarin and vanillin. The simplicity of the idea is truly amazing, using bold blocks of materials in a perfectly balanced accord. The place of vetiveryl acetate, as the main woody material of the earlier perfumes, has been taken by Vertofix (20%), with Galaxolide (20%) coming in as the dominant musks in combination with musk ketone, cyclopentadecanolide, and ethylene brassylate. These materials, together with vetiver, Hedione, and hydroxycitronellal make up some 75% of the formula.

The smell of Galaxolide has become very much associated with its use as a major component of many washing powders and fabric conditioners and its use here in conjunction with a high proportion of aldehydes may have been used deliberately to give the character of freshly washed white linen—or perhaps it was the fragrance that inspired the name.

The dominant floral character is again a fine quality rose note, using an unusually high proportion of nerol, reminiscent of Calandre. As previously noted in our description of Paris, DMBCA forms an important link between the rose notes and the woody materials. A fresh green character is introduced into the muguet aspect by the use of Liffarome backed up here by the more long-lasting acetaldehyde diphenylethylacetal.

The largely synthetic composition of the perfume again makes it important to use a natural rose oil, as part of the rose complex, in order to bring complexity and aesthetic quality. As in so many perfumes small amounts of carnation materials add depth and finish to the perfume. The typical floral aldehydic combination of aldehydes, floral notes, woody notes, and musk is completed by the addition of a powerful amber note.

THE FLORAL SWEET PERFUMES: L'ORIGAN, OSCAR DE LA RENTA, POISON, VANDERBILT

Lying somewhere in character between the florals and orientals are a group of perfumes whose origins go back to a style characteristic of the first decade of the century. At that time many new floral aromatic chemicals were in the process of being discovered. These were used for the creation of compounded floral bases, including such masterpieces as Jasmin 231, Dianthine, and Fleur D'Oranger—all introduced by the Swiss company Firmenich at around that time. These floral bases in turn were used by perfumers, many of whom worked independently from the supply houses, as the starting point and inspiration for the creation of new perfumes. In L'Origan, created by Coty in 1905, floral bases such as these were combined with methyl ionone, and an "ambreine" note based on vanillin, coumarin, and civet, with vetiveryl acetate, heliotropin, and the nitromusks as additional base notes. Together with a traditional top note of bergamot, orange, neroli, and ylang, this created a style of fragrance that was to influence the composition of such modern masterpieces as Oscar de la Renta, Vanderbilt, Poison, and Loulou. Many of these perfumes also include a mossy character similar to that of Mousse de Saxe, based on isobutyl quinoline.

Oscar de la Renta and Poison

Although a similar idea to L'Origan had been used in the creation of L'Heure Bleue (Guerlain 1912), few other surviving perfumes in the genre were created until the style was revived in 1976, by the introduction of Oscar de la Renta.

Despite the 70 odd years that separate the two perfumes Oscar de la Renta shows a remarkable similarity in its overall conception to that of L'Origan. Whether or not it was originally formulated, as was L'Origan, around the use of bases, we may conveniently describe its composition along the same lines. At the heart of the perfume is a similar combination of methyl ionone, an ambreine note, carnation, and orange blossom in the approximate proportions of 2:1:1:1. Heliotropin (7.5%), coumarin, and musk ketone add to the powdery sweet, semioriental character, with resins such as benzoin and opoponax. Vertofix (5%) is used rather than vetiveryl acetate as the main woody note, with a small amount of sandalwood.

Apart from carnation and orange blossom, the floral part of the perfume is dominated by a traditional jasmin complex to which has been added Hedione (8%) and Jessemal. Rose notes are also present as well as a small amount of hyacinth. The dominant sweetness of the perfume is lightened by a fresh top note made up of bergamot, orange, and mandarin, with a large amount of linalyl acetate (some 10% including that from the bergamot), and linalool. Ylang and basil, which contains a high proportion of estragol, lead in to the sweet floral and powdery notes.

Three other notes are important in completing the composition of this unusually complex perfume: orris, probably in the form of the resin, a fruity note built around the accord between phenoxyethyl isobutyrate and dimethyl benzyl carbinyl butyrate, and a mossy note reminiscent of Mousse de Saxe.

Two of the most important constituents of Oscar de la Renta, which give the perfume much of its character are the eugenol, contained in the carnation accord, and the Schiff bases (including methyl anthranilate) coming from the orange blossom. In Poison (Dior 1985) this combination was again used, with heliotropin, coumarin, and vanillin, but with even greater emphasis on the methyl anthranilate. This material, as well as the Schiff bases derived from it, forms the starting point for the creation of both orange blossom and tuberose notes, and in Poison it is the tuberose direction that has been taken, with the addition of aldehyde C18 (so-called), gamma-decalactone, and methyl salicylate.

Although best thought of in terms of its individual components rather than as a perfume constructed around the use of bases, the general structure of the perfume can clearly be seen to be related to that of Oscar de la Renta, the most obvious difference being the almost complete absence of methyl ionone. In its place the heavy floral character of the perfume has been reinforced by the introduction of a large amount of Lyral (10%). This material, being an aldehyde, also forms a Schiff base with the methyl anthranilate, as do two other floral aldehydes present in the formula, Canthoxal and Lilial. The musk element is greatly emphasized with Galaxolide (5%) and Celestolide reinforcing the musk ketone, and adding further to the overwhelming sweetness of the perfume.

In line with many other modern perfumes the fresh top note has been largely eliminated. It is replaced here by the somewhat narcotic effect of cypress oil, with significant amounts of rose oxide, and alpha and beta damascone adding to the fruity floral character—an effect that we also saw used in Paris. The combination of mandarin and geranium, which brings a welcome touch of freshness to the top note, occurs also in Giorgio, another predominantly tuberose perfume, though one belonging more to the floral salicylate family.

Much as we admire the technical excellence of Poison, we must admit to finding it a difficult perfume to live with—or to sit too close to in the theater or concert hall! For a perfume to intrude so ostentatiously upon other people's space can at times become irritatingly offensive.

In later perfumes, such as Loulou (Cacharel 1987), the oriental and tuberose notes are further developed along the lines of Oscar de la Renta to give what may be thought of as a hybrid or "floriental" effect.

Vanderbilt

Another perfume that may be included in the sweet floral family, and that has enjoyed great success in the United States, is Vanderbilt, created in 1981. Built around a combination of fresh citrus top notes, orange blossom, and tuberose, Hedione, methyl ionone, heliotropin, vanillin, and musk ketone, with iso E super coming in as the woody note, it lacks much of the heavy floral sweetness of the other perfumes in the group, lacking most of the eugenol with little or no ylang or coumarin.

Of particular interest here is the use of allyl amyl glycolate and allyl cyclohexyl propionate, combining to give a fruity green note, which may be thought of as replacing the fruit and hyacinth notes in Oscar. A honey note gives added character to the orange blossom aspect of the perfume. As we saw in White Linen, there is an admirable economy in the structure of the perfume, made remarkable by the subtle use of contrasting nuances.

THE ORIENTAL PERFUMES: SHALIMAR, MUST DE CARTIER, OBSESSION, YOUTH DEW, OPIUM, COCO

The oriental perfumes may conveniently be divided into two distinct groups. The first, which includes such perfumes as Shalimar, Must de Cartier, and Obsession is based on the classic "ambreine" accord between bergamot, vanillin (or ethyl vanillin), coumarin, and civet, usually with woody and rose notes. The second, to which Youth Dew, Opium, and Coco belong, has evolved around the relationship between benzyl salicylate and eugenol which we saw in L'Air du Temps, but here used in combination with patchouli and hydroxycitronellal (another important accord), spices, woody notes, and coumarin. This combination, which for convenience we will refer to as the "mellis" accord, had also been used as the basis for a number of important speciality bases such as Melysflor and Pimenal, which were used in the creation of such perfumes as Blue Grass and Moment Supreme.

All contain sweet balsamic notes—such as benzoin, tolu and opoponax—amber notes based on labdanum, and castoreum, with rose as the dominant floral.

Shalimar

Shalimar, created in 1925 by Guerlain, remains, in our opinion, one of the greatest and most aesthetically satisfying perfumes ever made. In structure it looks back to a style of perfumery reminiscent of the nineteenth century, based on a large proportion of expressed and distilled essential oils to which animalic and balsamic materials have been added as fixatives. Into this classical structure have been introduced two of the most important synthetic materials to have been discovered at around the turn of the century, ethyl vanillin (3%) and coumarin (9%). Ethyl vanillin is usually regarded as being approximately four times as strong as vanillin, so that it is not unusual to find as much as 10% or 12% of vanillin being used in perfumes of this type. The top note is dominated by some 30% of bergamot as well as other citrus oils, neroli, and rosewood. Making up the rest of the ambreine accord are patchouli (4%), sandalwood, civet, and vetiver.

Castoreum is of particular importance in Shalimar, forming the basis of the leather aspect of the perfume, which may be further developed by the use of a leather base, such as a classic Cuir de Russie. The same idea was used by Guerlain in Mitsouko, one of the earliest of the chypre perfumes to contain also a fruity note based on aldehyde C14 (undecalactone). Cinnamon bark oil, which forms part of the spicy aspect of the perfume, fits in perfectly with the leather. Other spices that work well are coriander, nutmeg, and clove.

The orientals in this group are some of the least floral of perfumes, relying mainly upon a small amount of rose, usually including both the natural oil and absolute, as well as a number of synthetics. In a perfume created at the time of Shalimar, the qualities of citronellol and geraniol used in the perfume would have been of natural origin, rhodinol from geranium, and geraniol from palmarosa. These products, because of the numerous trace materials retained from the starting material, have a quality and performance in the end product that is almost impossible to achieve using their purely synthetic counterparts.

At the time when Shalimar was made it was customary to use a high proportion of the alcoholic tinctures made from natural products, rather than the concentrated absolutes and resins. These would have included such products as vanilla, benzoin, and tonka to give naturalness to the ethyl vanillin and coumarin, as well as tinctures of civet, castoreum, ambergris, and musk. Such products retain more of the most volatile components of the starting material, giving a wonderful quality and life to the finished product.

Must de Cartier

Must de Cartier was created in 1981. The 70 years that separate it from Shalimar saw a revolution in perfumery style as well as in the availability of new synthetic materials. Although most of the structure of a classic oriental remains in place, to this has been added an accord based on approximately equal amounts of Hedione, sandalwood, and Galaxolide. These materials, together with vanillin, which replaces the ethyl vanillin in Shalimar, make up some 40% of the formula. Coumarin is again a major component, with cedryl acetate coming in to support the sandalwood. A small amount of a fruity-green accord based on Triplal and aldehyde C14 has been added to the top note. To this may also be added a trace of thyme—an idea that is again reminiscent of Mitsouko.

Obsession

A further development along these lines can be seen in Obsession (Calvin Klein 1985) in which the fruity-green and aromatic herbal notes have been emphasized even more, in a way reminiscent of Alliage (Lauder 1972). The classic combination of armoise, galbanum, and methyl chavicol (the major constituent of basil and estragon) that makes up part of this character is one that we shall see again in a chypre context, going back to perfumes such as Bandit and Cabochard. The use of tagete is particularly interesting for the natural, almost applelike nuance that it brings to the fruity-green character coming from small amounts of Triplal and allylamyl glycolate. Also contributing to this effect is Helional which acts as a link to the Hedione, used here in the same way as in Must de Cartier.

Otherwise, in general construction the perfume follows closely that of a classical oriental, with citrus oils (making up some 25%), lavender, sandalwood, patchouli, vanillin, coumarin, castoreum, and rose. The general sweetness of the perfume is further enhanced by the choice of musk materials, Galaxolide (5%), Tonalid, ethylene brassylate, and musk ketone, as well as by the use of Cashmeran, an intensely sweet fruity musk material associated with the synthesis of Galaxolide.

As in so many perfumes the placing of the amber notes is one of the most difficult tasks to accomplish well. Although many of these are based on ciste-labdanum, the total number and olfactory variety of such products is immense, and whether creating or working within the area of an established fragrance, the perfumer needs great skill and experience in their selection and use. On account of their relatively high boiling points the identification of labdanum products, as well as of other resins and balsamic materials, is largely beyond the range of GC analysis, and the student wishing to recreate the great perfumes of the past, which contain these materials, must again rely upon his or her purely olfactory skills.

Youth Dew

Youth Dew, originally launched by Estee Lauder in 1952 as a bath additive, is one of the few western fine fragrances not diluted in ethanol. Although as much criticized as admired for its sheer impact and lack of aesthetic subtlety, it remains one of the most original and influential of perfumes. Its enormous success, particularly in the United States and Great Britain, can be seen in retrospect to have opened the way for a demand for the style of perfumery that has now come to dominate the market.

The mellis accord, which makes up an important part of the fragrance here incorporates amyl salicylate in addition to benzyl salicylate, with the patchouli and woody aspect being accentuated, using cedryl acetate rather than the more traditional products derived from vetiver. Vertofix had yet to be discovered. The dominant spices, in combination with eugenol, are clove and cinnamon.

Although the mellis accord provides a framework for the structure of the perfume, it is the sweet balsamic materials, sandalwood, oakmoss, labdanum, vanillin, and animalic notes, in combination with the spices, that give the perfume its overwhelming character. Benzoin, styrax, tolu, and peru are all important, with castoreum and nitromusks contributing to the animalic aspect of the perfume. An animalic complex based on the esters of paracresol in combination with phenylacetates and cedarwood is probably also used.

Opium

Although the influence of Youth Dew can be seen in Dioressence (1970), it was not until 1977, with the launch of Opium by Yves St. Laurent, that the oriental theme was further developed in a perfume of major importance, which was in turn to provide the inspiration for the creation of a number of subsequent fragrances within the genre.

In Opium the two types of oriental, represented by Youth Dew and Shalimar, are brought together by the combination of the mellis and ambreine accords. Again there is an emphasis on castoreum, and on the spicy and balsamic notes, with the rose aspect being more fully developed than in either of its forebears.

In the top note, which in Shalimar was dominated by bergamot, there is a combination of citrus oils—orange, lemon, mandarin, and bergamot—together with linalool and linalyl acetate, and lavender. The typical mellis accord is built around a combination of benzyl salicylate (12%), eugenol with clove and pimento, patchouli (8%), hydroxycitronellal (10%), Vertofix, and coumarin (2.5%). The coumarin can be seen as acting as a link between the mellis and the ambreine accord, which is here based on a combination of both vanillin and ethyl vanillin. A combination of balsamic and resinous materials, including benzoin, styrax, opoponax, and tolu contribute to the overall oriental effect.

As we would expect in a perfume as distantly related in time as Opium is to both Shalimar and Youth Dew, a number of more modern materials have been introduced. The *cis*-3-hexenyl salicylate backs up the benzyl salicylate, and Lyral, the hydroxycitronellal. A small amount of Hedione is also included. There have been important developments to the rose aspect of the perfume with the addition of phenylethyl dimethyl carbinol (3%) and Centifolyl. The fresh character of the rose derives from the use of geranyl acetate. Unlike the earlier oriental perfumes there is also an important aldehydic note.

Coco

Coco, launched by Chanel in 1984, follows much the same pattern as Opium but with a greater emphasis on the floral notes. Hedione is here a major constituent, and Lyral has now replaced hydroxycitronellal as part of the mellis accord. These materials, together with the salicylates, eugenol, and patchouli, make up some 45% of the formula. Again we see the use of a strong structural accord at the heart of a modern perfume.

To the floral side of the perfume has been added ylang, bringing out the carnation aspect of the eugenol-salicylate accord. This, with the jasmin and rose components, gives a more floral effect reminiscent of some of the older mellis-based perfumes such as Moment Supreme. The presence of methyl ionone also tends in this direction. The woody side of the accord is predominantly a combination of Iso E super and sandalwood.

The renaissance of the oriental theme over the past 25 years can be seen as the starting point for a number of new directions in perfumery. One of these, which we saw in the creation of Must de Cartier, has been toward a group of dominantly sandalwood perfumes. Many valuable sandalwood synthetics are now available, including Brahmanol, Sandranol, Madranol, and Sandalore, and these synthetics have been used in conjunction with natural sandalwood in the creation of such perfumes as Joop and Samsara. Another important direction has been toward the so-called floriental perfumes, such as Loulou, which combine the sweetness of an oriental perfume with a floral complex usually based on tuberose. This type of perfume can be seen as a link between the true orientals and the sweet floral perfumes discussed in the previous section.

THE PATCHOULI FLORAL PERFUMES: DIORELLA, AROMATICS ELIXIR, CORIANDRE, KNOWING, PALOMA PICASSO

The ability of Hedione to combine successfully with many of the most important perfumery materials has been one of the major influences in perfumery over the past 30 years. Nowhere is this better seen than in a group of perfumes based on its relationship with patchouli. Although sometimes classified as chypres, the dominance of floral notes and patchouli, and the comparative lack of musk and animalic notes, other than castoreum, justifies the placing of them in a separate group.

Patchouli oil is one of the most widely used and valuable of perfumery materials, coming from the dried leaves of a plant that grows in Indonesia and other Southeast Asian countries. Most of the production is distilled locally, often in old-fashioned equipment from which the oil picks up a considerable amount of iron. As will be discussed in a later chapter, the presence of iron can have a disastrous effect on the stability of a perfume compound, and this needs to be removed either chemically or by means of redistillation in more modern equipment. Finer qualities of the oil are produced by the initial distillation being carried out under modern conditions, and large quantities of the dried leaves are imported into Europe for this purpose. A further rectification can also be carried out by molecular distillation, and it is these qualities produced in modern equipment that are most widely used in fine perfumery. A good quality patchouli has a character partly reminiscent of bitter chocolate and pepper.

Diorella

The original use of Hedione was in Eau Sauvage, created by Edmund Roudnitska in 1966, a predominantly citrus-fresh fragrance containing a number of aromatic herbs, such as thyme, basil, and armoise, with a background of jasmin, patchouli, Helional, eugenol, methyl ionone, woody notes, and coumarin. Many of these materials were again used in the creation of Diorella (Dior 1972), a unique and brilliantly conceived perfume, built around the accord between patchouli, Hedione, Helional, and eugenol.

The top note of Diorella is again dominated by lemon, verbena, bergamot, linalool, and linalyl acetate, with lavender and estragon. Apart from a small amount of rose the floral side of the perfume consists mainly of an intense jasmin note including Hedione (10%), *cis*-Jasmone (2%), hexyl cinnamic aldehyde, benzyl acetate, and indol, with aldehyde C14 providing the fruity aspect. As in Eau Sauvage, Helional (5%) and eugenol make an accord with the Hedione, carrying the essential character into the heart of the perfume in combination with the patchouli (6%), methyl ionone, vetiver, and a colorless oakmoss. The richness of the perfume comes from the use of jasmin absolute and trace amounts of other natural products. Petitgrain citronnier and cardamon both work well in an olfactory reconstruction of the perfume.

The "watery" character of Helional which plays an important part in the character of both Eau Sauvage and Diorella has subsequently been used both in the creation of cool fruity notes such as Calyx, and in modern "marine" notes based on Calone, such as New West and Escape.

Aromatics Elixir

In the same year that Diorella was launched another perfume, Aromatics Elixir, based on the combination of patchouli and Hedione, came on the market. Although in many ways quite different than Diorella, containing salicylates and with little of the citrus top note, many of the base notes such as methyl ionone and vetiver are again present with small amounts of eugenol, Lilial, and Helional. A further important difference is the very much greater emphasis on rose in combination with the jasmin. Hydroxycitronellal, remarkable for its absence in Diorella, is also present, forming another important accord with the patchouli.

This combination of patchouli, hedione, rose, hydroxycitronellal, methyl ionone, and woody notes has been used in a number of subsequent perfumes including Coriandre, Aramis 900, Paloma Picasso, and Knowing, which can be thought of as forming a small but characteristic family.

Coriandre and Knowing

In Coriandre (Couturier 1973) the top note is of coriander, ylang, styrallyl acetate, and undecylenic aldehyde, with geranium as part of the rose note. The type of rose base, which probably makes up some 10% of the formula, is a classic combination of phenylethyl alcohol, citronellol, geraniol, geranyl acetate, honey notes such as ethyl phenylacetate and phenylethyl phenylacetate, phenylacetaldehyde, rosatol, camomile, and violet leaf. Apart from hydroxycitronellal, Lyral, and phenylacetaldehyde glyceroacetal add to the muguet character.

Although we have described the perfumes in this family as being based on Hedione, in Coriandre the material used is Magnolione, a closely related material with a very similar though rather more jasminic olfactory character. This makes up some 20% of the formula with 10% of patchouli. Methyl ionone, vetiveryl acetate, cedryl acetate, Vertofix, and sandalwood make up the woody aspect of the fragrance, with a small amount of Galaxolide as the musk.

Knowing (Lauder 1988), although launched 15 years after Coriandre follows a very similar pattern, and the two perfumes are closely related in odor. In Knowing there is the addition of a fruity-tuberose character and heliotropin, with the woody side dominated by vetiver and a small amount of Iso E super. A number of other trace auxiliary notes have also been added.

Paloma Picasso

All the main elements that we have seen in the other perfumes in the group are again brought together in Paloma Picasso (Picasso 1984), though with the four major components, patchouli, Hedione, rose, and hydroxycitronellal in the approximate ratio of 2:1:2:2. On the same scale the main woody material, in this case Vertofix, is 2 and Lyral 1. Methyl ionone is again important, with castoreum providing the sweet leathery note. Of the perfumes discussed in this group Paloma Picasso is the most chypre in character having a pronounced mossy character.

What we frequently find in a perfume that shows descent through a number of previous perfumes is that the original idea, which may have been quite simple, has been greatly developed, both by the introduction of an increasing number of auxiliary notes, sometimes in the form of bases, and by single materials being replaced by complex mixtures. Paloma Picasso is a case in point, lacking perhaps some of the clarity of either Aromatics Elixir or Coriandre. Sometimes a simple reinterpretation of the original idea is needed to give new life to an old and overworked family.

THE CLASSICAL CHYPRE PERFUMES: MA GRIFFE, FEMME, MISS DIOR, CABOCHARD

The word "chypre" derives from the island of Cyprus, which for many centuries was the meeting point between East and West for the trade in aromatic materials. During the nineteenth century it became famed for the production of perfumes combining the citrus oils, floral pomades, and labdanum of the Mediterranean region, with resins and gums, such as styrax, incense, opoponax, and myrrh, imported from Arabia. Animal products such as civet from Ethiopia and musk from the Himalayas were also among the most valued commodities.

Originally, therefore, the word "chypre" would have been used to describe a style of perfumery associated with the island from which it came. Today, however, it refers rather more specifically to a group of perfumes whose origins can be traced back to the great Chypre of Coty created in 1917, a perfume based on oakmoss in combination with bergamot, jasmin, labdanum, and animal notes including civet and musk. To these may be added woody notes such as vetiveryl acetate, methyl ionone, and usually patchouli, to give the characteristic structure of most of the great chypre perfumes. Into this combination may also be introduced a variety of materials such as aldehydes, green notes, fruity notes, and leather, emphasizing different aspects of the chypre accord, to give a number of distinct subfamilies.

The great diversity of chypres has led some perfumers to classify perfumes under three headings: florals, orientals, and all the rest, which may be generally described as chypres. Certainly the widespread use of the materials that make up the typical chypre accord, in all types of perfume, gives some support to this view, and the nonperfumer, or even the perfumer, when faced with the sometimes unnerving task of commenting upon a new perfume for the first time, can usually get away with the description (unless the perfume is clearly an oriental) "floral with chypre aspects."

Today the classic chypre perfumes and their descendants are no longer much in fashion. In the United States consumer preference has moved in the direction of the sweet powdery floral notes, and in Europe, the former stronghold of the chypre style, the oriental and floriental notes have come to dominate the market. The trend in America has been linked to the widespread use of Johnson's Baby Powder, and the conditioning effect that this has had on consumer preferences. In European countries, where fresh Eau de Cologne type products are used in baby care, the oriental theme, which includes at least a vestige of fresh notes, has continued to be successful, while satisfying the demands of the all-important American market for warmth and sweetness. In Ysatis (Givenchy 1984), one of the few successful chypres of recent years, a more modern effect has been obtained by greatly developing both the jasmin and musk aspects of the accord with some 20% each of Hedione and Galaxolide. Indeed, with Evernyl being used in place of natural oakmoss, the overall effect is of a floral musk rather than of a true chypre.

Ma Griffe

To do justice to this, one of the most extraordinary of perfumes, we will break our self-imposed restriction on attribution by mentioning its creator, Jean Carles, by name, since the perfume reflects so completely every aspect of his art. Carles was also a talented conjuror who kept his pupils amused with the tricks he performed with cards as well as with those he performed with perfumery materials.

As we would expect the perfume is clearly composed around a classical structure of top, middle, and base notes. Apart from all the essential elements of a chypre there is the bold use of strong materials such as citronellal (1%), the aliphatic aldehydes (1%), styrallyl acetate (4%), and styrax in an accord that gives the perfume its enormous individuality. It was also part of Carles's technique to take structural materials such as vetiveryl acetate, methyl ionone, and hydroxycitronellal and dress them up with a number of natural and synthetic materials to produce bases that could then be used, rather than the individual materials, in the creation of his perfumes. The resultant complexity gave a wonderful velvety quality to the end product, which is difficult to duplicate by the normal processes of analysis and matching. It is probable that Ma Griffe was built up in this way using such bases as Selvone, Aldehone Alpha (based on hydroxycitronellal), and Muguet Invar, some of the great Roure specialities of that period.

Apart from the materials already mentioned the top note contains a simple mixture of bergamot and orange. The middle note is dominated by a traditional jasmin base, with muguet (hydroxycitronellal 10%), and rose. At the heart of the perfume is the classical chypre accord between vetiveryl acetate (10%), methyl ionone, oakmoss, aldehyde C14, coumarin, sandalwood, patchouli, musk ketone, and amber. The amber note can be reproduced by the simple but effective combination of labdanum, olibanum, and vanilla. Small amounts of other animalic notes such as civet may also be used.

The large amount of aliphatic aldehydes, together with the citronellal, make the perfume difficult to assess when first put into ethanol. A considerable period of maturation is needed for the aldehydes to "go in" with the formation of their hemi-acetals (see Chapter 18). Unfortunately, the use of citronellal in insecticides, and as an ingredient of many household cleansers, has given Ma Griffe a superficial association that is less than it deserves as one of the all-time great classics.

Femme

One of the earliest perfumes to use undecalactone (so-called aldehyde C14) was Mitsouko, created by Guerlain in 1919. This marvelous chypre still survives as one of the great artistic masterpieces of perfumery. As with many perfumes created around that time lilac played an important part in the floral side of the composition, and this, with the fruity notes and opoponax, was again used in the creation of Femme (Rochas 1942).

Reminiscent of an earlier style of formulation, Femme contains a relatively high proportion of fresh top notes based on bergamot and lemon, as well as petitgrain citronnier. At the heart of the underlying chypre accord is a fruity jasmin note (a jasmin benzol type may here be used), oakmoss, methyl ionone, patchouli, and labdanum, with small amounts of woody materials. The fruity note in Femme is probably introduced in the form of a base, such as the famous Prunol of de Laire, containing aldehyde C14. The underlying warmth comes from balsamic materials and from spices such as cumin and cardamon supported by a carnation based on eugenol, isoeugenol, heliotropin, and ylang.

Miss Dior

The versatility of the chypre type of formulation, in providing the starting point for the creation of perfumes of remarkably different character is nowhere better seen than in Miss Dior, created in 1947. Although the central chypre accord of bergamot, jasmin, oakmoss, patchouli, vetiveryl acetate, labdanum, and animalic notes comprises some 60% of the formula, the perfume, at the time of its launch, was one of startling originality. Even today it remains one of the most easily recognized, both when smelled on a smelling strip as well as in use.

Two years earlier Vent Vert had been created by Balmain, based on the use of galbanum oil, in combination with geranium, both at unusually high levels. (Sadly, the extreme character of this beautiful perfume has now been modified so as to appeal to a wider market.) This emphasis on green notes was again taken up in Miss Dior, combining galbanum with the accord between the aldehydes (C11 undecylenic and C10), styrallyl acetate, and a styrax note, found in Ma Griffe. In addition there is a dry spicy note based on pepper and coriander, as well as lavender, and neroli. This extraordinary balancing act between contrasting materials has always made Miss Dior one of the most admired perfumes among perfumers themselves.

A conventional jasmin base again provides the main floral aspect of the perfume. In more modern versions, for example, as in the Eau de Toilette, Lyral is used to replace part or all of the hydroxycitronellal. The styrax note may be reinforced by the use of phenylpropyl alcohol, one of its major constituents. The base note, in many ways similar to that of Ma Griffe, differs in the very much higher level of patchouli (10%).

Contrasting with the somewhat harsh top note of the perfume are warm amber and animalic notes, and the powdery softness of orris and vanillin. Natural jasmin and tuberose may be used to give richness to the perfume. A trace of celery seed oil also forms an interesting accord with tuberose.

At the time when Miss Dior was created most natural tuberose was obtained by the traditional enfleurage process (see Chapter 4). Today, partly because of high labor cost but also because of the religious restriction on the use of animal fats in producing countries such as

India, most tuberose absolutes are obtained by solvent extraction. The two types of product are very different in olfactory character, with the more modern product being greener and lacking much of the "jammy" richness of the original, which was so important in many of the earlier perfumes.

Apart from its influence on a number of other fine fragrances, Miss Dior types have found their way into many functional products, being particularly valuable in soap perfumery. The combination of green notes, lavandin, styrallyl acetate, and the aliphatic aldehydes; a floral character made up of ylang, fruity jasmin notes, and rose; PTBCHA, methyl ionone, and Vertofix; with a mossy amber background formed the basis of the famous Lux fragrance, created in the 1960s, and still the model for numerous subsequent brands.

Cabochard

In 1944, two or three years before the creation of Ma Griffe and Miss Dior, another very great and original chypre perfume, Bandit, had been launched by Piguet. Based on the use of isobutyl quinoline, this perfume was to become the inspiration for Cabochard, a softer and more commercial reworking of the same idea.

Although, in Cabochard, all the typical ingredients of a classical chypre are present, many important additions have been made. Yet despite the resulting complexity the perfume has a marvelous unity built around the various accords based on isobutyl quinoline.

Its somewhat earthy character blends perfectly with the top note combination of armoise (2%), galbanum, and basil. As in many modern perfumes containing green notes, mandarin is used as part of the fresh note, together with the traditional bergamot, linalool and linalyl acetate. Styrallyl acetate (2%), which in Bandit had formed part of a fully developed gardenia complex, is again important, and as in so many perfumes is combined with undecylenic aldehyde.

As would be expected in a chypre, the floral aspect of Cabochard is again dominated by jasmin and hydroxycitronellal (15%), but here with a small amount of hyacinth to back up the galbanum, and a fresh rose note given by geraniol. The jasmin is no longer the classic type found in Ma Griffe and Miss Dior but one based on benzyl acetate, Jessemal (4%), and hexyl cinnamic aldehyde (6%).

Compared with other types of chypre, Cabochard contains a relatively large amount of sandalwood (6%). Perhaps the use of this in combination with hexyl cinnamic aldehyde was, in another context, to influence the creation a year later of Madame Rochas. The same combination was again used in Aramis, a men's fragrance closely related to Cabochard, and one that was eventually to establish the quinoline-chypre accord firmly within the masculine area. Another important relationship used in the creation of both Cabochard and Aramis is that between isobutyl quinoline and musk ambrette. Now that this material can no longer be used, for reasons of safety, it is difficult to find an alternative musk that gives the same olfactory effect. Another musk used in Cabochard is ambrettolide, a very beautiful but expensive macrocyclic compound, which we also noted in connection with Fidji—a very different but not entirely unrelated perfume.

The typical chypre base note contains patchouli (6%), methyl ionone (the beta form was used in Aramis), vetiveryl acetate, cedryl acetate, oakmoss, and animalic notes. It is interesting also to find aldehyde C18 being used in a similar way to that of aldehyde C14 in many of the earlier chypres. An important part of the amber character comes from the use of Dynamone, a speciality material derived from cistus, difficult to use, but of great diffusion and persistence.

Other aspects of the perfume are its balsamic notes such as benzoin which, with castoreum and a costus note (the natural product can no longer be used), combine with the isobutyl quinoline to give the leather character. As we have seen before, spices such as cinnamon and clove work well with leather notes, and these are again present, backed up by a carnation accord of eugenol, heliotropin, and vanillin.

Cabochard is a perfume that repays a considerable amount of study, since it is full of interesting relationships between materials. Sadly, however, it is today more-or-less out of fashion, partly due the family having been hijacked by the male market but also because of its lack of many of the attributes required of a modern perfume. It is not monolithic in structure, sweet, powdery, or dominantly floral. Someday, when chypres again come back into fashion it may perhaps be reworked using modern materials such as Iso E super, Timberol, and Hedione, with perhaps a violet note, as in another of the older quinoline chypres, Jolie Madame.

CHANEL 19

Chanel 19, created in 1971, is the archetype of a small number of green perfumes that cannot be said to belong to any of the families or groups of perfumes that we have so far considered. Although all of its structural elements are among those that we have come across before, they are here brought together in an original way, to produce one of the great aesthetic masterpieces of traditional French perfumery—made even more remarkable for having been launched at a time when perfumery was moving more in the direction of lifestyle fragrances such as Rive Gauche and Charlie.

The perfume is perhaps as special for what it does not contain as for what it does. There are little or no aldehydes, synthetic musks, salicylates, patchouli, or vanillin. Although the floral and woody notes are reminiscent of some of the floral aldehydic perfumes, the green and mossy notes, together with the large amounts of orris and of Hedione, take it in an entirely different direction.

The top note of the perfume is dominated by galbanum with bergamot, lemon, and ylang. A trace of amyl acetate works well here in "topping off" the harshness of the galbanum. Rose notes including natural rose absolute, which make up some 15% of the composition, and a muguet complex (20%) similar to that used in Madame Rochas provide most of the traditional floral part of the perfume, together with a small amount of an "absolute" type jasmin speciality.

Apart from its green character the perfume is immediately characterized by the use of orris as a major component, rather than as a modifying note as in the majority of perfumes in which it occurs. Over the past few years the price of this already costly material has risen dramatically and to find the right quality for such a perfume becomes increasingly difficult. Although synthetic irone, its major ingredient exists, the performance of the natural material is hard to emulate. Methyl ionone (6%) provides a perfect link between the orris and woody materials.

The dominant woody aspect of the perfume is made up of Vertofix (12%), vetiver notes, cedryl acetate, and sandalwood, with guaiacwood supporting the rose. The mossy character comes from oakmoss and a complex based on isobutyl quinoline reminiscent of Mousse de Saxe. There is also a small amount of a spicy carnation note including one of the sweet spices such as clove or pimento. The wonderful richness of the perfume suggests the addition of numerous other materials in very small amounts, and it is probable that in the creation of a perfume such as this tinctures of musk, ambergris, and civet would also have been used.

Taking the top, middle, and base notes as we have described them so far, we have what can be thought of as a classically formulated perfume, similar in its structural proportions to perfumes such as Madame Rochas. But in addition to this, there has been added, as if by a stroke of perfumery genius, some 20–25% of Hedione, working throughout the perfume, giving life and diffusion to the whole, and a wonderful floral character. Smelled on its own Hedione seems to have little to recommend it. For this reason perfumers took some time to realize its true potential, but its ability to form a number of important olfactory accords and to raise the performance of a perfume has made it one of the most widely used materials today.

Unfortunately, few perfumes that are as green as Chanel 19 have a wide commercial appeal outside of the couturier market. However, one perfume that is clearly related to it and that has found some success is Silences by Jacomo (1979). Here the galbanum has been modified by the introduction of a cassis note to produce a perfume of outstanding individuality and beauty. So often we find this combination of green and fruity notes in successful modern perfumes. Most of the structure is very close to that of Chanel 19, though with the addition of benzyl salicylate giving a slightly more floral character.

Another perfume reputedly inspired by Chanel 19 is Beautiful of Estee Lauder (1986). Although much of the floral and woody aspects are certainly similar, the absence of green and mossy notes takes it into a different area. The addition of fruity and powdery notes to the remaining structure has produced a typically American perfume of considerable originality and character.

A NEW STYLE OF PERFUMERY: ETERNITY, TRÉSOR, SPELLBOUND, DUNE, CASMIR, AMARIGE

Over the past ten years the tendency toward the creation of monolithic perfumes, rather than those with a classic type of structure based on top, middle, and base notes, has seen the emergence of a new style of perfumery technique, which finally came of age with the creation of such perfumes as Trésor (Lancome 1990), Casmir (Chopard 1991), Dune (Dior 1991), and Spellbound (Lauder 1991). This technique has provided the inspiration—no doubt with the aid of information coming from GC analysis—for the launch, in quick succession, of a new generation of perfumes, which amounts to a revolution in the world of creative perfumery.

Although the initial stages in the progress toward this type of formulation were largely an American phenomenon, reflecting a consumer preference for strong, sweet, floral, and immediate smelling perfumes that can be relied upon to fulfill their initial promise in use—representing good value for money; its influence can now be seen even in the more traditional markets such as France and Spain.

Typical of the perfumes that immediately preceded this new wave of products is Eternity, a powdery sweet floral perfume built around a combination of benzyl salicylate and Iso E super, which together make up some 25% of the formula. In many ways the style of composition resembles that of a luxury soap fragrance, based on large amounts of comparatively inexpensive materials such as Lilial, Lyral, linalool, terpineol, eugenol, ionone beta, heliotropin, and Galaxolide. To these has been added a rich Osmanthus complex, possibly including the natural absolute, giving a fine floral and fruity character, together with the necessary feeling of quality.

The technique that has evolved to meet the demand for this new type of perfume revolves around the use of a comparatively small number of synthetic materials, all of which tend to be long-lasting and capable of being used at very high levels, of up to 25% or more, without the harshness associated with some of the older materials and naturals. Additional materials in the formula can be thought of as developing the character of the central accord, and providing decoration, rather than as being required partly to bring aesthetic harmony to an otherwise unrounded accord, such as in a conventional chypre based on oakmoss, vetiver, and patchouli.

Trésor is perhaps the most typical of the new type of formulation, being based on approximately equal amounts of Hedione, Iso E super, Galaxolide, and methyl ionone which make up some 80% of the formula. Around this massive construction, probably in the form of bases, are a combination of green, fruity, muguet, violet, and woody notes, many of them reminiscent of those used in Paris, together with heliotropin and vanillin. Precisely the same combination of four materials is again used in Spellbound, but here in combination with some of the carnation and tuberose aspects of Poison. Some slight relief to the overall sweetness is given by linalyl acetate.

In Casmir the same technique is used but in an oriental context. The formula is constructed around large amounts of Hedione (20%), Galaxolide and ethylene brassylate, a combination of Iso E super and Vertofix, and vanillin (14%), which together make up some 80% of the formula. In deference to the oriental character linalool and linalyl acetate are present to provide a top note with traces of rose, vetiver, and patchouli, in the background. A strong fruity note, characteristic of many modern American perfumes, is also included.

Another semioriental perfume dominated by the combination of Hedione and Galaxolide is Dune (Dior 1991). This may be seen as having some of the character of Obsession, with a fresh fruity-green top note, a combination of rose and jasmin notes in the middle, with sandalwood, patchouli, and vanillin (though less than in Obsession) in the base. Hedione and Galaxolide make up some 50% of the formula. The perfume is distinguished by a significant amount of orris.

A more traditionally floral perfume is Amarige (Givenchy 1991). Here the dominant accord is between Hedione (30%), benzyl salicylate (25%), and a combination of Iso E super and Trimofix. Around this structure are added a number of fruity notes, together with a rich floral background including ylang, tuberose, narcissus, jasmin, violet, and cassie.

Few of these perfumes fit easily into the generally accepted genealogy of perfumes; they represent a genuine break with the past. It is as if perfumers have discovered a whole new world in which to dream their dreams based on a new style of creation. Perhaps the future will see, as in the 1920s, a great flowering of new types of fragrance, based on this technique, with a few of the more successful being seen in retrospect as having been the inspiration behind whole new families of perfumes. The consumer will decide.

Part IV

Aspects of Creative Perfumery

13

Perfumes for Functional Products

The word "functional" in the title of this chapter may be thought of as redundant. One might argue that everything produced by humans, and certainly everything that is commercialized, has a function. Perfumers employ the expression "functional products" in a special way, that is, to denote all perfumed products other than alcoholic fragrances. The distinction is useful, and we will adopt it here.

In alcoholic fragrances, fragrance rules. The solvent system is part of the fragrance; the container and the delivery system are selected or designed to present the scent to its best advantage and to preserve it in its optimal state for as long a time as possible. In all other products, whether they be soaps or detergents, toiletries or household products, considerations other than optimizing fragrance performance take precedence. Soaps, shampoos, household cleansers, and detergents are formulated primarily to clean; in moisturizing creams and hair conditioners, the skin or hair care properties are more important than perfume performance. Even products such as bath foams and air fresheners, in which fragrance undoubtedly plays a central role, must meet requirements that may be in conflict with optimal fragrance performance. Bath foams must have a certain viscosity and must produce ample and stable foam; air fresheners must be inexpensive and easy to use.

As a result the creation of perfumes for "functional products" always involves considerations both of aesthetics (how should the product smell?) and of technique, of making the perfume appropriate to the product formulation or, as is often said, to the "product base."

In most perfumery textbooks, functional products have been dealt with on a product-by-product basis. Today the range of product types and product formulations that are perfumed has become so extensive and subject to so frequent changes as to make this approach impractical. We have therefore decided to deal with perfumes for functional products in a general manner, concentrating on principles of general validity. We will address the questions of fragrance aesthetics first and then review the technical considerations.

AESTHETIC CONSIDERATIONS

Masking

The bases of most functional products possess more-or-less undesirable odor qualities. In creams, soaps, and detergent-based products, it is the greasy or waxy note of fatty acids and their derivatives; in cold-wave lotions, depilatories, and insecticides, the pungent smell of active agents; in window cleaners and nail polish removers, the sweet "chemical" odor of organic solvents.

The severity of the problem depends a great deal upon the purity of the raw materials employed. The first and most fundamental step in eliminating off-odor problems is the use of the purest and least odorous grades of raw materials that are available and affordable. Where odor problems are caused by the breakdown of product components over time—such as the oxidation of unsaturated fatty acids which causes rancid off-notes, or the hydrolysis of the sequestering agent EDTA used in laundry detergents which results in a vinegar off-note—they are best eliminated by product reformulation or the selection of more suitable packaging materials.

Sometimes undesirable odors arise during or after a product's use: The wet laundry odor in washing machines or the hot hair scent generated by hair driers are typical examples. In personal deodorants and air fresheners, diaper products and toilet cleaners, masking external odors is a primary reason for the products' very existence.

Any odor can be covered by any other odor, provided that the covering odor is sufficiently powerful. Covering by drowning out may be a satisfactory approach in cases where the odor to be masked is faint. However, whenever this odor is distinctly noticeable, the drowning out is likely to result in an overall odor intensity that is in itself unpleasant. The art of masking is the art of *removing the unpleasantness of a given odor at the lowest possible total intensity level.*

A drowning-out perfume might be compared to the ritual masks, totally covering the face, worn by the actors at native ceremonies. An elegant masking perfume is like the small black masks worn at masked balls that cover only a small area around the eyes, just sufficient to make the wearer unrecognizable. The trick of masking at minimal intensity consists in using the odor to be masked as a component of the total odor or, to put this in another way, supplying a context within which the odor to be masked is no longer perceived as unpleasant.

Fatty odors, for example, are effectively masked by citrus notes, especially orange. This is so because the C_8 to C_{12} fatty alcohols and aldehydes, with their distinctly fatty character, are natural components of citrus oils. It is not by chance that cod liver oil, administered to children in former days as a rich source of oil-soluble vitamins, was commonly flavored with orange, and that citrus oils are major components of the Eaux de Cologne used to mask the fatty components of perspiration odor. Synthetic citrus oils in which the fatty aldehydes and alcohols have deliberately been left out are even more effective masking agents.

Perspiration often also has a pungent component. A similar odor is found as a component of neroli oil and, to a less extent, of oil of petitgrain. Both oils are common ingredients of citrus colognes, the former today in the form of synthetic replacements. These same oils, for similar reasons, are also effective in masking the characteristic odor of lanolin.

A good masking odor should be as similar as possible to the odor to be masked in its time-intensity curve. Fleeting off-odors can be masked by volatile masking agents, but a long-lasting off-note needs an equally tenacious masking note. Protein hydrolysate odors are so tenacious as to be difficult to mask over time.

Theoretically odors that actually counteract a given off-odor in the sense of decreasing its intensity, if possible to zero, would be even more ideal masking agents. In practice such annihilating odors have never yet been found. A few rare instances have been recorded where the addition of a second component results in reduced total intensity. But this phenomenon always involves a rather precise ratio of the two components, and it has as yet not been possible to put it to practical use.

Some components of product bases cause a stinging sensation in the nose that is not actually an odor but a response of the trigeminal nerve, a pain response. This sensation is faintly noticeable in poor grades of alcohol, more pronounced in acetic acid (vinegar), and very pronounced in ammonia and in formaldehyde. It cannot be masked in any way.

Sometimes off-odor problems appear more serious to the chemist who has developed the product than they are to the consumer. This is especially likely to occur in traditional products where the consumer through habituation has come to accept, and perhaps to expect, the base odor as part of the natural odor of the product. Such an example occurred in Germany in the late 1950s. For tax reasons, hair tonics and other aqueous alcoholic products had been formulated with isopropyl alcohol rather than ethyl alcohol. Isopropyl alcohol has a characteristic, somewhat sickly sweet odor. When the tax laws changed and manufacturers were able to use ethyl alcohol again, many consumers actually missed the familiar isopropyl alcohol note, and perfumers were asked, in some instances, to find ways of putting this note back into the products.

In window cleaners traditionally based on dilute ammonia, removing the ammonia sting does not necessarily lead to improved consumer acceptance. Consumers in some markets have come to expect the sting and regard it as a signal of product performance. On the other hand, cases have also occurred where off-odors that appeared insignificant to the experts caused a considerable drop in consumer acceptance.

A good general approach in the case of presumed malodor problems is to urge the client to conduct exploratory consumer research to find out whether the odor is truly perceived as objectionable by the consumer and, if so, at what stages of product use—for example, when opening the bottle, during product use, or after use. A clear definition of the problem, in consumer terms, places the perfumer in the best possible condition to solve it.

Evaluating the masking efficacy of a perfume is fairly straightforward if the odor to be masked occurs in the product base, although complications arise if the base odor changes markedly upon product aging. It is more difficult where the task of the perfume is to mask odors originating elsewhere, as in the case of deodorants, detergents, air fresheners, and toilet cleaners. Standard malodors simulating the odor to be masked are often used in evaluation tests conducted during the developmental stages. The definitive tests of masking effectiveness should involve actual in-use situations.

Pleasing

Whenever a functional product is perfumed, the main reason is quite simply to make it more pleasing to the consumer. A pleasant fragrance in a product is like the smile on a person's face: It is a signal of friendliness, an encouragement for closer contact, and one of the little joys that sweeten everyday life. As in the interaction between people, there are different stages in the encounter between the consumer and the product. The perfume comes into play in all of them. To illustrate this point, let us examine a shower foam.

Odor first comes into play during the shopping trip. The consumer has presumably decided beforehand that he or she needs some shower foam. If he (let us, for the sake of readability, take the proviso "or she" as given) has not yet firmly made up his mind which particular brand or variant to buy, opening a bottle that catches his attention and sniffing at the contents may well be the leading factor in this decision. The perfume must, at this stage, mask the odor of the plastic bottle and the fatty odor of the product base that have collected in the head space. It should in addition please the shopper both as an immediate sensation and as an indication of the odor to expect when he uses the product under the shower.

The second phase of the encounter takes place when the product, greatly diluted with water, is used under the shower. The user expects a pleasurable fragrance burst at this point, perhaps even a fragrance that pervades the bathroom.

Finally, there is the odor that lingers upon the skin after rinsing and drying. Depending upon the market for which the shower foam is intended and upon its advertising platform, a faint and rather neutral odor may be called for at this stage or a scent that is characteristic and distinctly noticeable.

This simple example points up the fact that in functional products, top, heart, and base notes of a perfume are not necessarily a continuum. They come to the fore during different phases of the product experience and play different roles in each. The demands are highly product specific. In a dishwashing detergent, for example, it is the top note that counts; no lingering afterodor must remain on the dishes. In a fabric softener, on the other hand, the odor remaining on the fabric after drying is essential. In an air freshener for prolonged use, top note, heart note, and base note should be as similar to each other as they can possibly be, both in character and in intensity.

The perfumer must not only be aware of consumer expectations at each phase of product evaluation, he or she must also know, as intimately as possible, the conditions under which the product will be used. Populations for whom toilet soap is the only regularly used fragranced product have quite different expectations regarding its ability to perfume the skin than consumers accustomed to daily use of alcoholic fragrances and deodorants. In countries where fabric softeners have high market penetration, the lingering power of a detergent perfume is less crucial than in countries where softeners are a rare luxury, and the demands placed upon a detergent perfume's heart note are different in societies where people wash their laundry in the river than in markets where washing machines are common. Space fragrances used in the non-air-conditioned taxicabs of tropical countries have different demands placed upon them than those used in the rooms of five-star hotels.

In the context of fragrance aesthetics of functional products, it is useful to divide these products into two categories. One group consist of products in which fragrance plays a major role from the consumer's point of view, such as deodorants, soaps, bath products, fabric softeners, and air fresheners. In the other group we find products such as shampoos and hair conditioners,* moisturizers and hand lotions, laundry detergents, and household cleaners in which fragrance is rarely seen by consumers as a key reason for using one brand rather than another, although it may actually have a considerable effect upon consumer acceptance. Fragrance here plays a more hidden, subliminal role.

Let us consider the latter category first. Although few people will consciously buy a product because of its odor, they may well reject it if they dislike its scent. Hence the tendency in these categories is to go for safe "nonpolarizing" fragrances, fragrances that may not be very exciting but that are disliked by few people. The safest fragrances are those that are reminiscent of the odor of the major established products in the market. These fragrances "stand for" high and reliable quality because the consumer has come to associate them with the category leaders that set the standard for quality in the market. Moreover they were probably selected in the first place because of their good performance in consumer tests. An extreme case of a category leader setting the fragrance standard for a category is the baby products market in the United States. For at least half a century, no product had a chance in this market unless its fragrance emulated, more or less closely, the traditional scent of the Johnson & Johnson baby line. There may nevertheless be reasons even in markets of this kind to depart from the established fragrances of market leaders. These will be discussed in Chapter 16.

*In Japan and other Pacific countries, fragrance is actually a major consideration in hair products. The role of fragrance in a given category may differ from market to market.

In the other group of products, where fragrance is a more overt reason for brand choice, odor types are a great deal more diverse and innovative. Here, a major force in fragrance evolution has come from alcoholic perfumery in what has been called the "trickle-down" phenomenon.

The Trickle-Down Phenomenon

There is a hierarchy among fragranced products, a hierarchy in terms of luxury and expense. Fine fragrances in limited distribution occupy the top position. In recent decades there has been a great deal of "trickling down" of fragrance types that were first introduced in fine fragrances into applications such as personal, laundry care, and even household products. We find Chloé-type fragrances in shampoos, Trésor types in fabric softeners, Drakkar Noir types in car air fresheners.

Turning to fine fragrances for inspiration has several advantages for the perfumer. Of all branches of perfumery, fine fragrance perfumery is most directly tied to fashion and life style. Innovations in tune with the spirit of the time first turn up here. Moreover it is often desirable to lend functional products an aura of luxury and elegance. What could be more effective in conveying such an aura than a note reminiscent of a popular fine fragrance?

Not all luxury fragrances are equally good candidates for trickle-down. Some are too polarizing, appealing strongly to a small group but finding little acceptance among the population at large. Some crucially depend upon materials that are technically unsuitable or unaffordable for many functional products.

Moreover trickle-down is often more apparent to the perfumer and to the product manager in the finished goods house than to the consumer. Few nontrained people recognize fragrances in their original form, let alone in the modified forms in which they are present in functional products. What comes across to the consumer is usually no more than a diffuse sense that the product smells feminine, fashionable, or stylish. For the perfumer using the trickle-down approach, it is therefore more essential to give a perfume optimal performance in the particular end product than to maintain maximum fidelity to the fine fragrance which was the source of inspiration.

We must qualify this last statement in the case of assignments that are related to but not identical with trickle-down: the line extensions of fine fragrances. When perfumes have to be developed for a soap, a body lotion, deodorant, or shower gel that are to be sold alongside a fine fragrance carrying the same name, closeness to the original becomes a far more important consideration, and compromises with respect to performance must sometimes be accepted for the sake of maintaining this closeness.

Both line extension assignments and the trickle-down approach involve the art of perfume adaptation, of replacing components of the original composition by others that are technically more suitable for the application in point, or less expensive. We will deal with the problems of adaptation in a later section of this chapter.

Communication

The observation that trickle-down fragrances often convey a sense of luxury and style points to an aspect of perfumes that is of key importance in the perfuming of functional products: The scent of a product connotes specific product qualities to the user. This is obvious when a lemon perfume is used in a dishwashing detergent advertised as containing lemon juice or in a sanitary cleaner based on citric acid, or when a herbaceous perfume is used in a herbal bath foam. But it also occurs in most if not all other uses of fragrance where the message is less specific.

Odors are a primary system of communication in the animal kingdom, and it is probably fair to say that humans also, consciously or not, interpret every odor as a communication about its source or carrier, as a signal of more or less specific characteristics (Jellinek 1991). Here again, we might compare the odor of a product to the expression on a person's face.

The product characteristic signaled by the fragrance depends a great deal upon the specific nature of the odor and upon the context. One type of lemon odor may suggest gentle freshness, another powerful cleaning action. A herbaceous accord may suggest active masculinity in a men's cologne and health benefits in a herbal bath.

Often the message is also culture dependent. The sweet powdery floral scent of Johnson & Johnson's baby powder conveys the protected safety of infancy to Americans, while it may well convey mature femininity in countries with a different scent tradition in baby products. Lavender in colognes is quite acceptable in Latin countries but considered old-fashioned in Germany.

The perfumer should generally beware of assuming that specific odor notes evoke the same associations for the public at large as they do for him or her. This is particularly but not exclusively true for the perfumer in an international fragrance house working on projects for different markets. Consumer research, designed and interpreted with

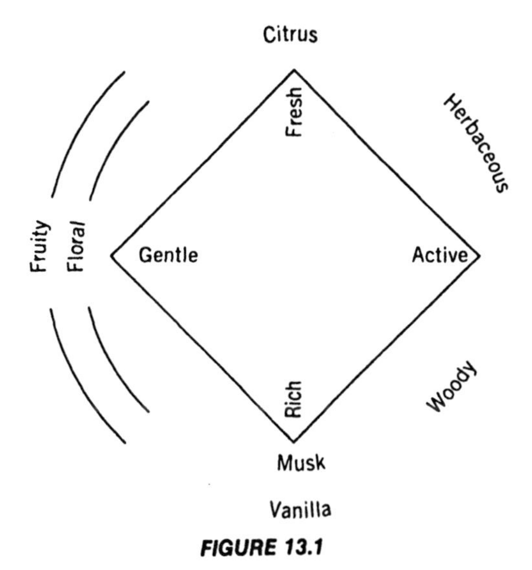

FIGURE 13.1

understanding, can yield a great deal of information about what qualities people associate with what odors. We strongly advise the perfumer to use market research information of this kind if available.

The odor effects diagram of Paul Jellinek (1954) gives useful general guidance in this area. It was developed for fine fragrances, but in Figure 13.1 we have adapted it for functional products. Jellinek's diagram was based both on psychological reasoning and on empirical experience, and it was validated, many years after its publication, in a number of consumer studies (Jellinek 1994). For the original labels "erogenic" and "antierogenic" we have substituted "rich"and "fresh," using the latter term despite the fact that perfumers often find it difficult to comprehend what it means to consumers, simply because we know of no more suitable word. The original "narcotic" and "stimulating" we have replaced by "gentle" and "active."

The floral and fruity notes span a wide range from cool, fresh accords such as violet, geranium, apple, and cassis to rich ones such as orange blossom, tuberose, peach, and raisin. For the sake of clarity, not the entire spectrum of possible notes is shown in the diagram. Eucalyptus, pine, and camphoraceous notes belong in the upper right fresh-active segment; aromatic and spicy notes are on the far right, active side; mossy, leathery, and amber notes are on the lower right, active tending to rich; civet and honey notes are at the lower left, rich tending toward gentle; the fatty aldehydes are on the upper left, fresh-gentle. Most perfumes contain components of several or all of the classes shown in the diagram. A perfume's dominant message is that of its predominating components.

TECHNICAL CONSIDERATIONS

Performance

The term "performance" is generally used to denote a perfume's ability to make its presence noticed. It is crucial in the perfuming of functional products because the objective is always, for both economic and technical reasons, to achieve maximum odor effect at the lowest possible perfume dosage.

The question whether the performance of a perfume depends only upon the performance of its individual components, or whether the art of composition is also a major contributor to performance, has been little explored even though it really is basic to perfumery. In any case the skill to forge harmonious accords from individually highly performing materials is an essential part of modern perfumery technique.

Depending on time and location, different aspects of performance can be distinguished:

	Near the Source	At Some Distance
Soon after application	Impact	Diffusion
After some time	Tenacity	Volume

Impact refers to the efficacy of a perfume during the first moments of product experience, for example, when sniffing at the bottle or applying the product to the skin. **Diffusion** is a measure of the distance over which the fragrance is noticeable soon after application. High diffusion is desirable, for example, in the case of a bath foam or a dishwashing detergent where the consumer often looks for an immediate burst of scent.

Tenacity refers to the long-term effectiveness of the fragrance in the perfumed product (as in a soap bar or an extended action room freshener) or on the surface to which the product has been applied, for example, upon the skin after use of a toilet soap. In former days it was believed that tenacity dependent upon the use of so-called fixatives (evaporation-retarding agents) in composing a perfume, today the opinion pervails that it is simply the resultant of the tenacity of the individual odorants used (Jellinek 1978).

Substantivity refers to the ability of a perfume or a perfume material, applied in a diluted dispersion in water, to attach itself to a solid surface such as the skin (in the case of toilet soap and bath and shower products), the hair (in shampoos and conditioners), or textile fibers (in detergents and fabric softeners). It may refer also to their ability to stay on the surface when this is moistened (as in deodorants or waterproof sun protection products).

Volume is the effectiveness over distance, some time after application. The difference between tenacity and volume often lies at the root of misunderstandings between the perfumer and the client. The perfumer, accustomed to examining fragrances at close range, measures the lasting power of his or her creations by their tenacity; the consumer and the marketing manager look for volume.

The Three Approaches

In the evaluation of performance, three approaches are possible: empirical, semiempirical, and systematic.

The empirical approach involves testing each perfume in the product base in which it is to be used and observing all relevant aspects of its performance during the essential phases of product use. The empirical approach makes no assumptions and answers precisely the crucial question: How does *this* perfume perform in *this* base?

This is the approach evaluation boards use when selecting perfumes for submission. It can, however, give perfumers only limited guidance in their creative efforts, since it consists in testing after a perfume has been completed. The empirical approach must therefore be supplemented by the *semiempirical* approach. Here the perfume materials are tested individually in the product base, again checking for all aspects of their performance. The assumption is that perfumes composed largely or exclusively of materials that perform well in the product base will also perform well.

The tests on individual perfumery materials are usually run prior to any specific projects. Even where they are conducted in the context of a specific project, they must be completed at an early stage of the project, when the exact composition of the product base is usually still in doubt. Consequently they are normally not run using the exact formulation of the finished product but a "standard formulation" that is typical of an entire finished product category. Therefore the semiempirical approach usually involves a second assumption of which we know that it is rarely entirely true: the assumption that the perfume will perform in each specific finished product formulation in the same way in which it performed in the standard formulation. Obviously it is important that the "standard formulations" used in the tests be as realistic as possible.

Considering the perfumer's palette of several thousand perfume materials and the large and ever-increasing number of formulations of functional products of all kinds, pretesting all materials in all bases is an enormous, costly, and time-consuming undertaking. The magnitude of the effort is multiplied by the fact that several panel members usually take part in the evaluation, since judgments of odor performance are to a certain extent subjective.

In the late 1950s several perfumers and perfume chemists proposed that fragrance performance, in all media and applications, is largely dependent upon the rates of evaporation of the individual perfume materials under the specific conditions in which the fragrance is evaluated (Pickthall 1956; Sfiras and Demeilliers 1956; Jellinek 1959). They stressed the importance of knowing the vapor pressures of perfume materials (Appell 1964). They suggested that differences in the performance of a given odorant in different media are caused by the attraction forces between odorant and product base, and that these forces could be understood and predicted from the chemical structures of the materials involved (Jellinek 1961; Dervichian 1961).

They proposed a systematic approach to perfumery, an approach that draws upon general laws and observations relating the various aspects of performance of odorants to their physical and psychophysical properties and that relates these in turn to their chemical structure. If our knowledge of these relationships were complete, we could dispense with laborious empirical testing altogether. But even with incomplete knowledge, the systematic approach can, in theory, be of great service in limiting the effort of empirical testing by focusing it upon the materials most likely to give positive results.

Although the techniques of vapor-phase gas chromatography were still rudimentary at the time, it was recognized that this technique could provide essential data for the study of odorant performance in solution (Jellinek 1961), in soaps (Sfiras and Demeilliers 1964, 1966, 1974), in talcum powders and detergents (Schiftan and Feinsinger 1964), and on human skin (Jellinek 1964). These early suggestions were carried further in the 1970s by Roehl and Knollmann (1970), Saunders (1973), and Burrell (1974).

Unfortunately, the complexities at all levels (physical, psychophysical, and chemical) are enormous, and our abilities for prediction of performance are accordingly limited and fragmentary. This limitation, coupled with the fact that the necessary data are laborious to gather and hard to find in the literature, has discouraged many perfumers from considering the systematic approach as a useful aid to their practical work.

We believe, however, that the time is approaching when the perfumery profession will pay more attention to this approach. Today the rapid development of chromatographic techniques has enabled data to be more easily obtained than in the past. The measurement of odorant concentration in the vapor phase of the perfume compound, of the perfumed finished product, and even of treated products such as wet and dry textile fibers is now becoming a standard technique (Neuner-Jehle and Etzweiler 1991; Müller, Neuner-Jehle, and Etzweiler 1993), and water solubilities of highly insoluble perfume materials can also be more easily established than in the past.

To be sure, the economic pressures on the perfumery industry are becoming such that we can no longer afford not to use all approaches that can make our work more efficient, even those that are, like the systematic prediction of performance, far from perfect at present. We will therefore briefly present this approach here, outlining its principles and giving some practical examples in this chapter, and providing basic theoretical background in Chapters 16 through 18.

The Basic Systematic Approach

To get some preliminary idea of how a certain perfume material is likely to perform, we should know its volatility and its solubility in water.

Measures of Volatility The **volatility** of a perfume material determines its staying power. The more volatile materials are the ones that evaporate more quickly; they tend to dominate the top note of a perfume and to be evident, for example, when sniffing at a freshly opened bottle or jar of a perfumed product or when examining the material on a freshly dipped blotter. The least volatile materials are the ones that stay around, even on fabrics after washing with a perfumed detergent; they are most noticeable in the base note of a complex perfume. Limonene and benzyl acetate, for example, have high volatility; synthetic musks and benzyl salicylate have low volatility.

Perfumers determine volatility by observing the behavior of perfume materials on the smelling blotter. All of the traditional classifications of volatility (Jellinek 1954; Poucher 1955; Carles 1961), and at least one more recent one (Sturm and Mansfeld 1975), are based on this approach. However, the process of evaporation from a paper strip involves complex physical interactions between the paper and the perfume material, and hence the paper strip is not a good model for physically different systems such as a soap, a body lotion, or even the human skin.

A more general measure of volatility, which has the added advantage of not being dependent upon individual judgments of intensity, is the vapor pressure at normal ambient temperature. The vapor pressures at 20°C of some common aroma chemicals are given in Tables 13.1 through 13.3. In Table 13.2 the materials are listed in order of increasing vapor pressure, and this is particularly useful for our purposes.

At this point, some critical remarks are in order. The tables are short, and many important materials are not listed. Absent are all the naturals, and 20°C is a rather arbitrarily selected temperature: at the skin surface 30°C would be more appropriate. What is more, compared to such readings as "it persists on the blotter for more than two weeks" which has some recognizable meaning, the values of vapor pressure, such as 0.00008 micron, are very abstract. Let us briefly address these objections.

The shortness of the list is due to the paucity of published data and to the fact that establishing vapor pressure data requires painstaking experimentation. However, once the value of the data is generally recognized it will not be long before more complete listings will be available within individual companies and hopefully also in the literature.

The naturals are omitted because they are mixtures of different components, each with its distinct vapor pressure. The vapor pressure of an individual component is the parameter of interest, and not that of a mixture. The problem with naturals in the classical blotter observation test is that their odor character changes during the observation, and not just their intensity.

Of course, the relevant temperatures will differ according to the circumstances: 20°–25°C is the customary room temperature for sniffing at products in the jar or bottle; it is about 30°C for sniffing at the skin surface, about 35°C for the bath, and 60°C or higher when opening a washing machine. Although the *absolute* values of vapor pressures are quite different at, say, 35°C than at 20°C, we are here concerned only with *relative* values, and these do not change significantly over the range from 20° to 35°C. For products subjected to washing machine temperatures of 60°C, special calculations would be necessary.

Since we are concerned with relative rather than absolute values, the actual vapor pressures are rather meaningless. The perfumer does not have to know precisely what 0.12 for vanillin means, but should

TABLE 13.1 Vapor Pressure and Water Solubility of Some Odorants in Alphabetical Order

Substance	Vapor Pressure (microns at 20°C)[a]	Water Solubility (ppm)[b]
Aldehyde C11	77.40000	20
Aldehyde C12 lauric	48.00000	20
Aldehyde C12 MNA		1
Ambroxan		1
Amyl cinnamic aldehyde	0.76000	40
Amyl salicylate	3.00000	6
Benzaldehyde	800.00000	3300
Benzyl acetate	87.00000	1700
Benzyl salicylate	0.08600	1
Cedrol		15
Cinnamic alcohol	13.00000	6000
Citronellol	163.00000	450
Coumarin	3.40000	1900
Cyclopentadecanolid	0.25000	<1
Decalactone, gamma	5.35000	750
Ethyl vanillin	0.07700	2900
Eugenol	9.85000	2200
Hexyl cinnamic aldehyde		10
Indole	9.20000	
Ionone, alpha	10.00000	190
Isoeugenol	2.90000	1400
Lilial		50
Limonene	1540.00000	50
Linalool	230.00000	2000
Linalyl acetate	68.00000	160
Lyral	19.60000	7400
Maltol	0.00008	12,000
Methyl anthranilate	10.60000	2700
Methyl ionone		50
Methyl ionone, gamma		60
Musk ketone		<1
Musk xylol	0.00534	<1
Phenylacetaldehyde	270.00000	2600
Phenylacetic acid	2.90000	18,000
Phenylethyl alcohol	36.00000	20,000
Phenylpropyl alcohol	15.00000	9900
Pinene, alpha	2900.00000	2
Terpineol, alpha	23.00000	2900
Tonalid	0.00050	<1
Vanillin	0.12000	10,000
Vertofix coeur		10

[a]Obtained by extrapolation from published vapor pressures at other temperatures.

[b]Obtained by gas chromatography of saturated aqueous solutions of the substances, using benzyl alcohol as an internal standard.

TABLE 13.2 Vapor Pressure and Water Solubility of Some Odorants in Order of Increasing Vapor Pressure

Substance	Vapor Pressure (micron at 20°C)[a]	Water Solubility (ppm)[b]
Maltol	0.00008	12,000
Tonalid	0.00050	<1
Musk ketone	0.00100	<1
Musk xylol	0.00534	<1
Ethyl vanillin	0.07700	2900
Benzyl salicylate	0.08600	1
Vanillin	0.12000	10,000
Cyclopentadecanolid	0.25000	<1
Amyl cinnamic aldehyde	0.76000	40
Isoeugenol	2.90000	1400
Phenyl acetic acid	2.90000	18,000
Amyl salicylate	3.00000	6
Coumarin	3.40000	1900
Decalactone, gamma	5.35000	750
Indole	9.20000	2900
Eugenol	9.85000	2200
Ionone, alpha	10.00000	190
Methyl anthranilate	10.60000	2700
Cinnamic alcohol	13.00000	6000
Phenyl propyl alcohol	15.00000	9900
Lyral	19.60000	7400
Terpineol, alpha	23.00000	2900
Phenyl ethyl alcohol	36.00000	20,000
Aldehyde C12 Lauric	48.00000	50
Linalyl acetate	68.00000	160
Aldehyde C11	77.40000	20
Benzyl acetate	87.00000	1700
Citronellol	163.00000	450
Linalool	230.00000	2000
Phenylacetaldehyde	270.00000	2600
Benzaldehyde	800.00000	3300
Limonene	1540.00000	50
Pinene, alpha	2900.00000	2

[a]Obtained by extrapolation from published vapor pressures at other temperatures.
[b]Obtained by gas chromatography of saturated aqueous solutions of the substances, using benzyl alcohol as an internal standard.

observe that this value is a great deal larger than 0.00008 (maltol) and a great deal smaller than 3.4 (coumarin).

Practical Implications of Vapor Pressures

As Jean Carles (1961) already knew, the best perfumery accords are those involving materials of comparable vapor pressure.

TABLE 13.3 Vapor Pressure and Water Solubility of Some Odorants in Order of Increasing Water Solubility

Substance	Vapor Pressure (micron at 20°C)[a]	Water Solubility (ppm)[b]
Musk ketone	0.00100	<1
Musk xylol	0.00534	<1
Cyclopentadecanolid	0.25000	<1
Tonalid	0.00050	<1
Ambroxan		1
Aldehyde C12 MNA		1
Benzyl salicylate	0.08600	1
Pinene, alpha	2900.00000	2
Amyl salicylate	3.00000	6
Vertofix coeur		10
Hexyl cinnamic aldehyde		10
Cedrol		15
Aldehyde C11	77.40000	20
Aldehyde C12 lauric	48.00000	20
Amyl cinnamic aldehyde	0.76000	40
Lilial		50
Methyl ionone		50
Limonene	1540.00000	50
Methyl ionone, gamma		60
Linalyl acetate	68.00000	160
Ionone, alpha	10.00000	190
Citronellol	163.00000	450
Decalactone, gamma	5.35000	750
Isoeugenol	2.90000	1400
Benzyl acetate	87.00000	1700
Coumarin	3.40000	1900
Linalool	230.00000	2000
Eugenol	9.85000	2200
Phenylacetaldehyde	270.00000	2600
Methyl anthranilate	10.60000	2700
Ethyl vanillin	0.07700	2900
Terpineol, alpha	23.00000	2900
Indole	9.20000	2900
Benzaldehyde	800.00000	3300
Cinnamic alcohol	13.00000	6000
Lyral	19.60000	7400
Phenylpropyl alcohol	15.00000	9900
Vanillin	0.12000	10,000
Maltol	0.00008	12,000
Phenylacetic acid	2.90000	18,000
Phenylethyl alcohol	36.00000	20,000

[a]Obtained by extrapolation from published vapor pressures at other temperatures.
[b]Obtained by gas chromatography of saturated aqueous solutions of the substances, using benzyl alcohol as an internal standard.

Maltol and benzaldehyde are examples of materials that are difficult to use in perfumes because their vapor pressures are outside the range of most common materials.

In a study of the stability of perfume materials in soaps and laundry detergents, Burrell (1974) found that loss by evaporation was a major cause of odor change.

In studies of the substantivity of nine aroma chemicals in a fabric softener application, head-space gas chromatography showed that benzyl salicylate, aldehyde C12 MNA, lilial, musk ketone, and the synthetic musk Fixolide were lost to a far lesser extent upon drying than were alpha ionone, eugenol, or linalool (Müller, Neuner-Jehle, and Etzweiler, 1992) (Figure 13.2). This result corresponds to the relative vapor pressures of these materials, except for aldehyde C12 MNA which is a special case.

Practical Implications of Water Solubility

The water solubility of a perfume material comes into play in a number of ways. For one thing, highly water-soluble materials tend to perform disappointingly in water-based products such as moisturizing gels or fabric softeners. This is a pity because they are exactly the ones that give the least problems of solubility in such bases.

Perfume materials of relatively high water solubility also appear not to be very effective when applied to water-friendly surfaces; this is noticeable, for example, when fabric softener and detergent perfumes are examined upon cotton fabric after drying (Müller et al. 1992). In addition they tend to be lost as fabric, skin, or hair is rinsed with water. In oil-based products, on the other hand, and on more oil-friendly surfaces they may perform well.

Water solubility is of course also an essential predictor of the clarity of perfume solutions, be they alcohol-water based or oil based and, as will be explained below, it also plays an important role in the stability of odorants in different media.

Actually what comes into play in perfume applications is not so much the solubility of an odorant in water as its distribution between water and other media or surfaces, for instance, between water and air (in the surface over a fabric softener solution), between water and oil (in a cream or in a rinse-off sitution on the skin surface), or between water and fabric fibers (in the use of a fabric softener or a detergent).

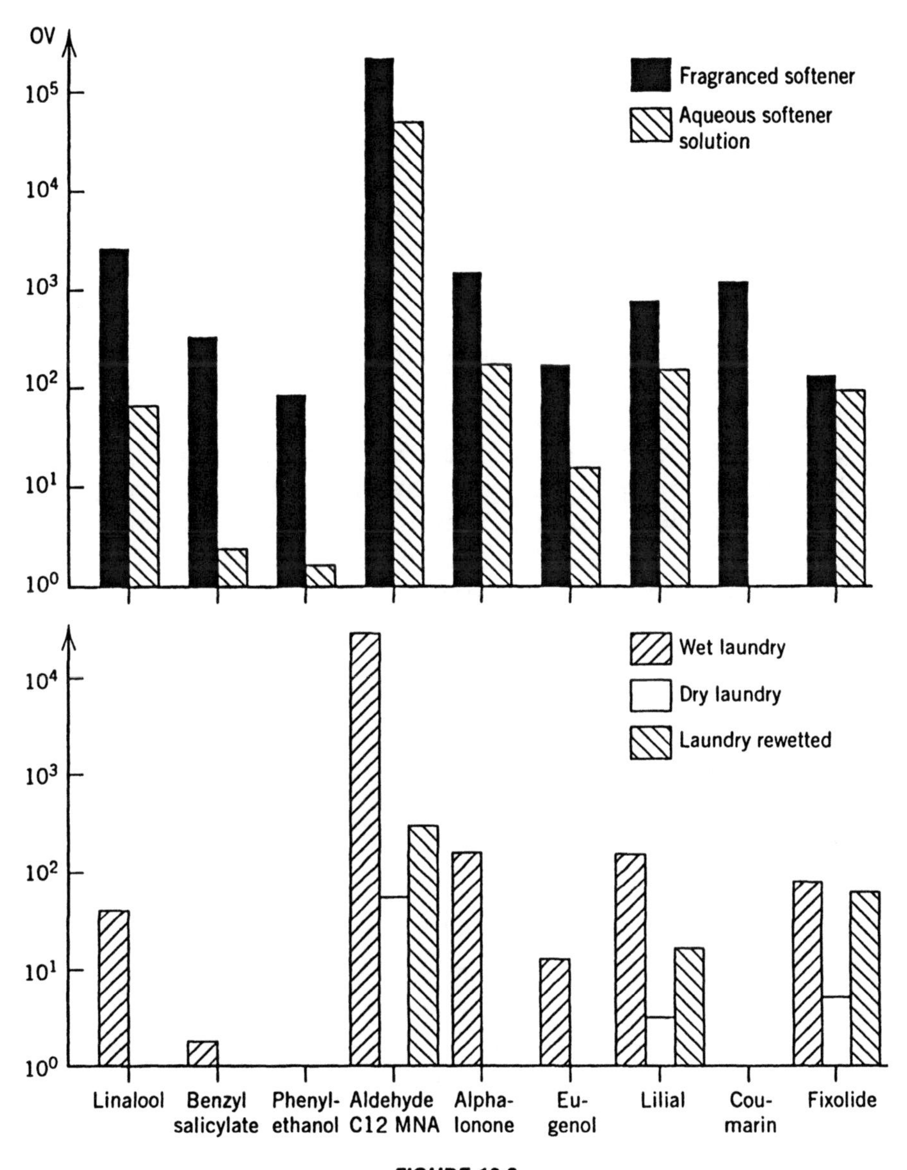

FIGURE 13.2

However, water solubility is a reasonably good indicator of such distributions in most cases.

In the above-mentioned study of the substantivity of odorants in a fabric softener (Müller et al. 1992) the practical consequences of water solubility were clearly apparent (Figure 13.2). The three materials with high water solubility (phenylethyl alcohol, coumarin, and benzyl acetate) lost their odor effectiveness nearly entirely as the fabric softener was dissolved in the rinse water, and they contributed practically no odor to the laundry. All of the other odorants remained clearly noticeable at least on the wet laundry. Along the same line, an earlier investigation (Jellinek and Warnecke 1976) had shown that amyl cinnamic aldehyde was adsorbed four times more effectively on a polyester-cotton (65:35) fabric than the far more water-soluble cinnamic aldehyde.

The water solubilities of selected odorants are given in Tables 13.1–13.3, with Table 13.3 presenting the odorants in order of increasing solubility. Table 13.4 (page 164) indicates for the most important functional products whether perfume materials or high or low vapor pressure and water solubility are more advantageous in each particular case, considering both impact upon examination of the perfumed product in its container and retention after product application. In addition special chemical and other considerations relevant to the applications are indicated.

Taken together, Tables 13.1 through 13.4 represent a step toward a systematic approach in the perfuming of functional products. Let us emphasize once again that the data should be used only as general *indications of the likelihood of satisfactory performance.* Because of the complexity of the systems involved, they must be always checked by empirical observation.

Inherent Odor Strength, Odor Value and Odor Volume

Next to volatility and water solubility, two additional parameters of odorants greatly affect their performance in application: inherent odor strength and odor volume. They are not included in Table 13.4 because they are nearly always desirable. Odor volume may be more or less relevant in different applications, but it is hardly ever undesirable (except in the residual odor of dishwashing detergents); inherent odor strength may be considered to be the basic measure of an odorant's value in terms of intensity. They were not included in Table 13.4 also because few reliable data have been published regarding either of them. Since they depend on the reproducible measurement of human responses, their determination is laborious and fraught with pitfalls.

What **inherent odor strength** means can perhaps best be explained by an example. When first examining a perfume of the Shalimar family on a blotter or on the skin, the odor impression is dominated by the citrus oils. Soon, however, components such as vanillin and ambergris and musk odorants begin to dominate the picture. At this stage their actual concentration in the head space is still considerably less than that of the limonene, linalool, and other components of the citrus oils, but they dominate because they have greater inherent strength.

The theoretical intensity of an odorant under any specific set of conditions can be roughly expressed in terms of its odor value under those conditions, which is defined as

$$OV = \frac{\text{Actual head-space concentration of odorant}}{\text{Threshold concentration of odorant in air}}$$

Odor values are only approximate relative measures of intensity and performance; the reasons for this are explained in Chapter 20. A more realistic but still rough measure of intensity is the adjusted odor value OV' which we define as

$$OV' = OV^{0.35}$$

We can define the inherent odor strength of a perfumery material as its adjusted odor value in its undiluted state at room temperature. Table 13.5 provides such values for a few selected materials.

An illustration of the use of odor values as measures of performance in a specific application is given in Fig. 13.2. The ten odorants were all present at the same level (3.6%) in the model mixture that was incorporated in the fabric softener. However, their odor values in the head space over the fabric softener differ, from the most effective (Aldehyde C12 MNA) to the least effective (phenylethyl alcohol), by a factor of several thousand. The differences are even more pronounced in the wet, the dry, and the rewetted laundry.

As a rough indication of relative cost effectiveness in a given application, the "cost per unit of adjusted odor value" may be calculated. Table 13.6 shows these costs for the ten odorants in the example shown in Figure 13.2, both for the odor over the fabric softener, which comes into play when a prospective buyer sniffs at the product before deciding whether to buy it, and for the headspace over the rewetted laundry,

TABLE 13.4 Perfumery Materials for Functional Products

Product Category	Stage[a]	Desirable Physical Characteristics		Special Considerations
		Vapor Pressure	Water Solubility	
Skin Products				
Deodorants	Application	+	0	In sticks, watch for discoloration
	LT	–	–	
Antiperspirants	Application	+	0	Must be stable to acids
	LT	–	–	
Creams, lotions (emulsified)	Application	+	0	
	LT	–	–[c]	
Gels (aqueous)	Application	+	–[d]	May need solubilizers
	LT	–	–	
Lotions (aqueous-alcoholic)	Application	+	–[d]	May need solubilizers in low-alcoholic formulations
	LT	–	–	
Skin, massage oils	Application	+	+	Select solvents for oil solubility (e.g., BB, IPM)
	LT	–	–	
Powders	Application	–	0	Must be stable to air oxygen
	LT	–	0	
Depilatories	Application	+	–	Must be stable to thioglycollate and alkali
	LT	–	+	
Lipstick	Application	+	+	
	(LT)	+	+	
Toilet soap, beauty bars	Application	+	–	Watch for discoloration
	LT	–	–	
Foam bath, shower foam	Application	+	–	If acidified to pH $\leq$ 5, esters may be hydrolyzed
	LT	–	–	
Bath salts	Application	+	–	Package must be impermeable to water vapor
	LT	–	–	

Hair Products				
Shampoos, rinses	Application	+	–	If acidified to pH ≤ 5, esters may be hydrolyzed
	LT	–	–	
Setting foams, lotions, sprays, gels	Application	+	–	
	LT	–	0	
Cold wave, hair straighteners	Application	+	–	Stable to thioglycollate and alkali
	LT	–	–	
Laundry Products				
Detergents, fabric softeners	During washing	+	–	Dependent upon specific formulation
	On wet laundry	+	–	
	On dry laundry	–	–	
Household Products				
Hard surface cleaners	Application	+	–	
	LT	–	–	
Dishwashing liquids	Application	+	–	
	(LT)	+	0	
Automatic dishwashing detergent	Application	+	–	Stable to alkali, peroxides (air)
	(LT)	+	0	
Scouring powders	Application	+	–	Stable to air
	LT	–	–	
Air freshener, wick type	LT	±[e]	–[f]	
Air freshener, gel type	LT	±[e]	?	
Air freshener, spray type	Application	+	0	

[a]Long-term fragrance performance (LT) is relevant only where desired. Products where residual fragrance is *undesirable* are marked by "LT"

[b]"+" means that a high value of the characteristic in question is desirable for optimal performance at this stage, "–" means that a low value is desirable, and "0" means that the characteristic is irrelevant to performance at this stage.

[c]Particularly relevant if the treated skin is likely to come in contact with water, such as in hand lotions and sun care products.

[d]High hydrophylicity (water solubility) favors solution but reduces performance.

[e]Volatility must be neither too high nor too low. It should be roughly equal for all components of the fragrance.

[f]Low substantivity to the wick fabric is important.

TABLE 13.5 Odor Values of Selected Substances at 20°C

Substance	CAS-Number	OV	OV'
Phenylacetic acid	103-82-2	3,000	16
Citral	5392-40-5	8,000	23
Bornyl acetate	5655-61-8	10,000	25
Styrallyl acetate	10522-41-5	10,000	25
Cinnamic aldehyde	104-55-2	30,000	37
Citronellal	106-23-0	60,000	47
Aldehyde C10	112-31-2	200,000	72
Aldehyde C11 undecylenic	112-45-8	200,000	72
Nonadienal	6750-03-4	3,000,000	185

Note: The vapor pressures used in the calculation of these values were obtained by extrapolation from literature data on boiling points at various pressures. The thresholds were taken from Devos et al. (1990).

which roughly represents what may be perceived when the laundered and dried garment is being ironed or worn. The data show that Aldehyde C_{12} MNA, the synthetic musk Fixolide, and Lilial, although they are among the most expensive of the set in terms of $/kg, are the most cost effective on the rewetted laundry. All the others, having essentially no odor effect at this stage, have an infinitely high effective cost.

Odor volume at first glance appears to be similar to inherent strength, but there is a difference. Let us again think of an example. When comparing undiluted benzyl acetate and alpha ionone on blotters, benzyl acetate appears to be the stronger one; yet when using the two in perfume formulations, low levels of alpha ionone usually make a great deal more difference, and are more effective than comparable levels of benzyl acetate. Benzyl acetate *loses its effectiveness more quickly upon dilution* than does alpha ionone; it has a lower odor volume. Materials that combine high odor volume with high inherent

TABLE 13.6 Cost Effectiveness of Selected Materials in Fabric Softener

	Price	OV' units		$/OV' unit	
Substance	$/kg	FS	LR	FS	LR
Linalool	10.00	16	0	0.60	∞
Benzyl acetate	2.00	8	0	0.25	∞
Phenylethanol	6.00	50	0	0.12	∞
Aldehyde C12 MNA	18.00	60	6	0.30	3.00
alpha-Ionone	38.00	12	0	3.20	∞
Eugenol	5.00	5	0	1.00	∞
Lilial	14.00	11	2	1.30	7.00
Coumarin	12.00	12	0	1.00	∞
Fixolide	22.00	5	5	4.50	4.50

Note: FS = fragranced softener, LR = laundry rewetted.

strength are diffusive, in that they can be smelled at a great distance and are very effective at the dry-out stage. In Chapter 20 odor value will be explained more fully.

Stability

Modern branded products must be unvarying in their quality. At each new purchase the consumer expects to find a product that is indistinguishable from the one bought the last time. Translated into a specification for perfumes, this means that the perfume must be stable in the product or, to put it more precisely, it must undergo no noticeable change, not in odor character nor in intensity nor in color, during the entire shelf life of the product.

This operational definition contains two qualifications. We do not say that the perfume must not undergo any changes whatever but only that these changes must not be noticeable to discerning regular users of the product. Nor must the odor be changeless forever, but only during the period after production during which the consumer is likely to find the product on a retail store shelf, plus some reasonable time of use thereafter. Since a manufactured product does not usually appear on retail shelves until some two weeks after its production, the changes in odor character and intensity that are often inevitable during the "maturing" period of the product do not present a problem, provided the odor has stabilized after at most two weeks. How long the normal shelf life of the product is at the far end depends upon the distribution channels through which it is sold. Today we usually deal with maximum shelf lives of 18 to 24 months.

In this section we will consider in a general way some factors that affect the stability of perfume materials in products. Since most changes in odor during product storage are due to *chemical interactions* between perfume materials and components of the product base, we will look at these interactions first. *Color changes* will be covered in a separate subsection. Finally, we will look briefly at perfume changes caused by *physical* factors.

Chemical Stability The chemical stability of any perfume material in any base depends upon its chemical structure and the composition of the base. General observations about the chemical stability of perfume materials as related to their chemical structure are to be found in Chapter 16.

The stability of many perfume materials is impaired in product bases whose pH lies outside the neutral range of about 6 to 8. Esters are especially affected: They tend to be split (hydrolyzed) into their constituent acids and alcohols. This leads not only to a gradual loss of their characteristic note, it also causes, in products with pH below 6, the odor of the acid to appear. Thus acetates may generate a vinegar off-note, and butyrates a rancid butter note. The degree of instability differs widely among esters. Streschnak (1991) mentions allyl amyl glycolate and linalyl acetate as being particularly prone to hydrolyze and suggests replacing the latter by terpinyl acetate in difficult media. Burrell (1974) found all formates to be extremely unstable in soap; methyl heptine carbonate also exhibited limited stability. Among the acetates, benzyl acetate showed poorest retention in soap but here, the problem may be one of volatility. Among the aldehydes, phenylacetaldehyde and benzaldehyde had extremely poor stability; hydratropic aldehyde was also rather poor; citral and cinnamic aldehyde were intermediate; and amyl cinnamic aldehyde, hydroxycitronellal, and undecylenic aldehyde showed good stability. There is here an indication of an inverse relationship between stability and water solubility; only hydroxycitronellal which has a rather high solubility is an exception. Acetals are also unstable in acid media, being split into the aldehyde and the alcohol from which they were derived.

Among the product bases whose pH usually lies outside the neutral range are acid and alkaline household cleaners; cold wave lotions, hair straighteners, and depilatories (alkaline, $pH > 10$); and antiperspirants (acid, $pH < 5$). Fabric softeners are acidic, with pH values between 3 and 5. Machine dishwashing detergent powders are alkaline and contain perborates, which are oxidizing agents. Bath salt tablets are also usually alkaline. The so-called neutral pH skin care products are usually acidified to pH 5 to 5.5 (the natural pH of the skin) and may therefore cause some hydrolysis of esters.

Oxidation is another rather frequent cause of perfume instability. It may be caused by oxidizing agents that are part of the product formulation (e.g., the hypochlorites in bleaches and in automatic dishwasher detergents, sodium perborate or percarbonate in heavy duty detergents and in laundry pretreatment powders) or by air in the case of powders such as talcum or scouring powder, where the perfume is distributed over the large surface of the powder particles. The unsaturated terpenes contained in citrus oils and in pine needle and other coniferous oils are particularly likely to undergo air oxidation, leading to a turpentinelike odor. The oxidizing agents contained in laundry and household cleaners also attack other perfume materials, especially those containing double bonds.

Oxidation is also the main cause of perfume instability in poor-quality toilet soaps; here it is caused by the fatty acid peroxides formed at the stage of incipient rancidity. Perfumes may to some extent be protected against oxidation by the incorporation of 1% to 2% of an antioxidant such as BHT (butylated hydroxy toluene) into the perfume oil.

The enzymes contained in some laundry detergents and prespotting agents may also cause perfume stability problems. They usually include esterases that destabilize esters and oxidases that attack double bonds. This, however, is not normally a serious situation, the more since in modern detergents enzymes are usually incorporated in encapsulated form.

The reactions that cause the breakdown of perfume components all take place in the aqueous phase of products. In dry products such as powdered dishwashing detergents, scouring powders and bath salt tablets, this means that perfume stability problems are much reduced as long as the products remain absolutely dry. Therefore packaging that effectively prevents absorption of moisture from the air can help a great deal in preventing perfume breakdown, and poor packaging can be a major contributor to perfume instability.

For the same reason the addition of an oily phase to water-based products can help considerably in promoting perfume stability. Most perfume materials are far more readily soluble in fats and in oils than in water, and to the extent that the product formulation provides an oil or fat phase into which they can escape, they are sheltered from the attack by the hydroxy ions of bases, the hydrogen ions of acids, the oxidizing agents and enzymes—all of which operate only in the water phase. Hence cloudy cold wave lotions, which contain emulsified oil, are easier to perfume than clear ones, cream depilatories cause less severe problems than fat-free formulations, and so on.

This effect of physically shielding unstable perfume components from contact with the water phase may perhaps be provided also by the stable components of the perfume compound itself, provided that the latter are present in sufficient excess. This would account for Paul Jellinek's observation (1954, p. 116) that "the presence of a soap-resistant aromatic delays the decomposition of compounds which are unstable to alkali, and may prevent it altogether," and for the common observation that unstable but powerful perfumery materials such as methyl heptine and octine carbonates which are normally used at very low levels exhibit better stability in practice than tests of their stability in pure form would have indicated.

Color Stability In addition to instability manifested by odor changes, there is also perfume instability that results in color changes. These are particularly bothersome in light-colored soap bars but may also cause problems in products such as creams and lotions, deodorant sticks, and shampoos. Usually they involve a progressive darkening from colorless through yellow to brown tones, although instances of reddish discolorations are also known (e.g., by indole). The most important perfume materials that often cause discoloration in soap are listed in Chapter 18; they may cause problems in other media as well.

One cause of such color changes is the reaction of aldehydic perfume materials with amines that are present in the perfume compound itself (usually in the form of methyl anthranilate) or in the product base. The more common cause, however, is the complex formation of a phenolic perfume material with Fe^{2+} (iron) or other metal ions present in the base or in the perfume oil itself. Patchouli oil is a frequent cause of iron in perfume oils but oak moss, cananga oil, or in fact any phenol-containing essential oil kept in iron containers may also be involved. Geranium oil may be rich in copper ions, which may also cause discolorations. Since the metal-phenol interactions occur in the water phase, they take place less rapidly in formulations containing an oil phase in which the perfume is nearly entirely dissolved and thus shielded. This effect explains Streschnak's (1991) observation that discoloration problems are less severe in superfatted than in nonsuperfatted soaps. The addition of chelating agents such as citric acid or EDTA, which bind metal ions, to the water phase can also alleviate the problem. The problem of soap discoloration by perfumes has been thoroughly discussed by Pickthall (1974).

Discoloration is also sometimes observed in alcoholic perfumes in clear glass bottles. Here the problem involves the action of light and air upon sensitive perfume materials. It can sometimes be alleviated by the addition of a sunscreen to the finished perfume. Sunscreens should not be incorporated directly in perfume oils since they may contribute to instability.

The Need for Testing

The majority of the reactions leading to perfume instability are known and understood. Hence the degree of stability of a given perfume in a given medium is, in theory, predictable. In practice, however, a great deal of hard-to-foresee factors may intervene due to the chemical and physical complexity of perfumes and of product bases. Therefore consideration of the chemistry involved is helpful in pinpointing likely points of trouble, but no firm predictions of stability can usually be made on the basis of theoretical considerations.

Practical aging tests of the finished perfume in the finished product base must be conducted to ascertain perfume stability in each concrete case. These tests are normally conducted at elevated temperatures (40°–45°C), since chemical reactions take place more quickly at higher temperatures. A practical rule of thumb is that the rate of change doubles with every 10° rise in temperature. However, stability tests are rarely run at temperatures exceeding 50°C, since too drastic deviations from normal storage conditions cause distortions in the predictions of stability.

Other Causes of Odor and Color Changes

Although the perfume oil is usually the first suspect whenever odor or color changes occur in a finished product, it is not always the culprit. Odor and color changes in the product base itself may occur due to oxidation, hydrolytic breakdown, complex formation, bacterial decomposition, or other causes. Sometimes the causes for instability are hard to track down, as in a case in the experience of one of the authors, where an off-odor in a cream was due to microbial breakdown that was made possible by absorption and inactivation of the preservative by the plastic container. It is always advisable to conduct a stability test of the unperfumed product along with the test of the perfumed product.

Compatibility

Compatibility with the Product Compatibility is, in a sense, the reverse side of stability. While stability is a question of the effects of the medium on the perfume, compatibility has to do with the effects of the perfume upon the medium. Ideally the perfume has no effect whatsoever upon the product base. Special care in perfume formulation may be required to achieve or approach this ideal. In products containing inherently unstable oxidizing agents such as hypochlorites or peroxides, perfume components may accelerate the breakdown of these agents and thus deactivate the product. Dyes used in coloring products may be decolorized or otherwise affected by perfume materials.

These are examples of chemical effects of perfume materials. Most cases of perfume incompatibility, however, involve physical effects. The most common one is the clouding of clear products, caused by insufficient solubility of the perfume in the medium. This is most likely to occur in aqueous-alcoholic products, especially where alcohol grades of 75% or lower are used. The problem is aggravated at low temperatures. Since materials, once they have dropped out of solution, often do not readily dissolve upon warming, problems of clouding are most likely to occur during cold weather if the perfumed product is stored out of doors.

The culprits often but not always are resinoids, essential oils that contain high levels of terpene hydrocarbons such as citrus and conifer oils, and high melting crystalline perfume materials. In borderline cases the problem may be eliminated by replacing such materials by more readily soluble substitutes (e.g., resinoids by absolutes, the essential oils by their deterpenated counterparts) and by replacing any phthalates or other solvents the perfume oil contains by dipropylene glycol. The problem should be attacked, in addition, by adjusting the method of preparation of the final product. The perfume compound should first be dissolved in the alcohol; then the water should be added gradually, with vigorous stirring.

Whenever such remedies do not suffice and the solubility of the perfume in the medium is insufficient to provide the desired odor intensity, recourse must be sought in **solubilization** of the perfume. This is achieved by blending the perfume oil with a suitable surface active agent (surfactant). Nonionic surface active agents are usually used for this purpose; ethoxylated fatty alcohols based on C12 to C18 alcohols and seven to nine ethoxy units, available under a variety of trade names in different countries, have proved very useful, as have surfactants based on ethoxylated castor oil. The amount of solubilizer required depends on the nature both of the product base and of the perfume; any component of the base that aids perfume solution (e.g., ethyl alcohol) and any perfume components that have high water solubility (see Table 13.3) lower the amount of solubilizer needed by raising the proportion of the perfume that goes into actual solution and does not need to be solubilized. For the same reason dipropylene glycol should be used as the solvent in the perfume formulation rather than phthalates or other solvents with low water solubility. A ratio of surface active agent to perfume of 3 to 1 often suffices to provide clear solutions. By fine-tuning a specific formulation, testing different surface active agent blends and different surfactant-to-perfume ratios, lower ratios can usually be achieved. This is advantageous both in terms of cost and of perfume effectiveness (Bell 1985) but is the application chemist's rather than the perfumer's task.

Another form of clouding that may be caused indirectly by the perfume is sometimes observed in water-detergent systems such as foam baths, shampoos, or dishwashing detergents. This, paradoxically, often becomes apparent at elevated temperatures and tends to disappear upon cooling.

Occasionally perfume oils may also cause clouding in water-free, oil-based formulations such as bath oils or massage oils, especially if these contain high proportions of mineral oils. The problem can usually be eliminated by limiting the levels of resinoids, high-melting crystalline perfume components and perfume materials with very high water solubility in the perfume formulation and by replacing any glycols or phthalates that it may contain by isopropyl myristate or another branched-chain fatty acid ester.

Perfumes may give rise to changes in the viscosity of detergent–water-based systems such as foam baths and shower foams, shampoos and liquid detergents. This is particularly apt to occur in formulations that contain relatively high levels of water and that are thickened by the addition of salt. The changes may be toward higher or lower viscosity, they are very much dependent upon the exact product formulation, and they are hard to predict. Sometimes water-miscible solvents such as dipropylene glycol, used in the perfume oil, are the cause of a lowering of viscosity.

Occasionally perfumes may cause destabilization of emulsions. This is an indication of inherent borderline stability of the emulsion system and should be remedied, wherever possible, by a change in the product formulation. Where this is out of the question, the only remedy is a laborious trial-and-error reformulation or replacement of the perfume compound. If high levels of phenylethyl alcohol or benzyl alcohol occur in the perfume, they may be the cause of the problem.

Compatibility with the Container In the experiments by Burrell (1974) with superfatted soap and laundry detergent it was apparent that although chemical instability was observed in some instances, a far more general cause of odor loss in products whose packaging is odor permeable is the evaporation of perfume components. The vapor pressures given in Tables 13.1–13.3 can serve as a general indicator of the risk of loss by evaporation.

Burrell actually measured the retention of perfume materials applied to a commercial detergent powder (Persil Automatic) in cardboard containers. Table 13.7 is extracted from his paper. With all of the materials examined, loss in a glass container over the same periods at room temperature was nil, indicating that there was no loss due to chemical breakdown. The odorant loss from cardboard was much greater at elevated temperature (37°C/70% rh). The rather inconsistent results comparing the loss after 12 and 24 weeks are hard to explain.

TABLE 13.7 Loss of Perfume Materials From a Detergent Powder after Storage at Room Temperature in a Cardboard Container

Material	Percentage of Material Present after	
	12 Weeks	24 Weeks
alpha-Ionone	86	78
Dihydrojasmone	92	72
Phenylethyl alcohol	66	52
Citronellyl acetate	65	31
Linalool	38	27
Phenylethyl amyl ether	35	23
Benzyl acetate	18	13
Benzyl amyl ether	40	12
Linalyl acetate	32	12
Phenylacetaldehyde	20	7
Citral (*cis* and *trans*)	40	0

Source: Burrell (1974).

Odor loss by diffusion of perfume materials can also occur in products packaged in plastic containers. Its rate depends on the following factors:

1. *The thickness of the plastic.* The diffusion rate is inversely proportional to the plastic's thickness.
2. *The nature of the plastic.* The relative diffusion rates through low-density polyethylene (LDPE) to high-density polyethylene (HDPE) to polypropylene (PP) are roughly as 100:10:1. For polyvinyl chloride (PVC), the diffusion rate may be 100–200 times less than for polypropylene.
3. *The nature of the product base.* Here the relationship is complex. On the one hand, diffusion is far greater from nonpolar solvents such as hydrocarbons and, presumably, fatty acid esters than from polar solvents such as alcohol because the nonpolar solvents cause the plastic to swell, thus increasing its permeability. On the other hand, product bases such as oil-free aqueous systems tend to favor the migration of perfume materials with very low water solubility into the plastic.
4. *The nature of the perfume material.* Molecules with a large molecular diameter diffuse far less readily than smaller ones, and polar materials (with relatively high water solubility) less readily than nonpolar ones (compare Table 13.8).

Although odor-impermeable packaging materials are usually preferable since they prevent perfume loss, a certain amount of package

TABLE 13.8 Diffusion Constants of Some Perfume Materials Through High-Density Polyethylene (HDPE) from a Methanol Solution at 23°C

Substance	Diffusion Constant (*D*) in $cm^2/s \times 10^{10}$
Camphor	0.22
Citronellol	0.35
Dimethyl benzyl carbinol	0.45
Menthol	0.57
Linalyl acetate	0.82
Eugenol	1.30
Phenylethyl alcohol	2.30
Diphenyl methane	3.50
Diphenyl oxide	3.90
Limonene	5.70
cis-3-Hexenol	15.00

Source: Becker, Koszinowsky, and Piringer (1983).

permeability to small volatile molecules may actually be helpful if the base generates off-odors due to small molecules (i.e., acetic acid). For this reason packaging materials that permit "breathing" are greatly preferable to completely impermeable ones for soaps and talcum powders.

Interactions between the perfume oil and the plastic are possible. Fortunately the instances of such interactions are rare, but when they occur, most likely in water-based, oil-free products with high perfume levels such as gel or liquid air fresheners or rim blocks, they can be quite annoying, since they may lead to the leaking of containers, the clogging of valves, and so on. Esters are particularly likely to interact with the more commonly used plastics. The problem may sometimes be remedied by replacing any esters used as solvents in the perfume compound by dipropylene glycol and by formulating the perfume to provide the desired intensity at the lowest possible perfume level. Discoloration of white plastic is usually caused by the same perfume materials that cause discoloration in white soaps (see Chapter 16).

Compatibility with the Encapsulation Process The encapsulation of perfumes normally involves a coacervation process in which water is the external phase. High levels of components with high water solubility cause problems. The use of water-miscible solvents such as dipropylene glycol must be avoided.

Adaptation

Creative assignments such as trickle-down exercises or developing fragrances for line extensions involve the adaptation of a fragrance formulation that had been developed for one type of product (usually alcoholic products) for use in other product categories. Adaptation may be thought of as a two-step process. In the first step, components of the formulation that are not suitable for the new base are replaced by similar-smelling and more suitable materials. In the second step, the formulation obtained in this way is rebalanced and optimized for the new medium.

The primary reasons for the replacement of components are chemical instability, discoloration problems, problems of fragrance incompatibility with the product base (e.g., solubility problems), problems of incompatibility with the packaging, and demands of costing. The subsequent necessity to rebalance the formulation may also arise from several reasons:

1. The materials used to replace the original ones may have greater or lesser intensity than the original ones.
2. The relative performance of the different components may change greatly as a result of the switch to a different product base (e.g., from an oil-free to an oil-containing product; Jellinek, 1959; Saunders, 1973).
3. The aesthetic requirements of the new application may demand greater or lesser emphasis on the top note or the base note.

Examples of case 3 are the adaptation of an alcoholic fragrance for use in body lotions, skin creams, or room fragrances. In all three cases both top and base note should be deemphasized in favor of the heart note. In an extended use room fragrance, in particular, the range of volatility of the components should be reduced as much as possible in order to achieve constancy of odor character.

Additional adjustments may be necessary in the case of product bases that pose masking problems. Adaptation assignments rarely represent challenges of creativity, but they may place considerable demands upon the perfumer's skill.

14

The Challenge of New Materials

For a perfumer a rare and great challenge is to be confronted with a new material, one that has as yet no known history of perfumery use. It may be a newly synthesized aroma chemical or a material derived from a not commercially cultivated plant.

The perfumer using a known material is like someone promenading in a well-tended park. The paths are clearly marked, one knows where they lead. The scope that is left to one's imagination lies in wandering off the beaten path, exploring a thicket off to the right or finding, in a certain season and under certain conditions of light, an unexpected view of rare beauty.

The perfumer faced with an entirely new material is like a scout confronting a virgin forest. The chemical structure (or, in a natural material, its main components), and whatever information the perfumer has regarding the material's occurrence in nature serve, like a compass and a crude survey map, to give him or her some direction, but there are no paths and there is no assurance of finding anything but weeds and chaos. It takes a brave person to venture in, a skillful one to detect the obscure intimations of passages, a lucky one to stumble upon spots of beauty, and a creative one to recognize the potential value of such spots. Only a perfumer endowed with a sense of adventure and with dogged persistence has a chance of being successful, of clearing a path or two and of emerging from the forest with a map that he or she can pass on to others.

Not all new materials are entirely new. Many are, in their odor character, reminiscent of familiar ones; the very first perfumer who smelled Lyral, for example, must have recognized its similarity to hydroxycitronellal. Others resemble familiar ones not only in their odor but also in the chemical structure; *cis*-3-hexenyl salicylate was bound to find its place, if it had any at all, within the general area of the earlier known salicylates.

With such materials, the perfumer's path of exploration is marked in advance. The perfumer must find out whether the new material possesses any significant positive differences compared to the familiar ones, differences that make the material worthy of being added to the standard repertoire; these may be differences in odor qualities such as character, tenacity, or volume, or advantages in application characteristics such as stability, cost, or skin compatibility.

No matter whether a new material is radically novel or reminiscent of familiar ones, its exploration requires imagination and an excellent gift of observation, and it is beset by uncertainties: Will its odor quality be precisely reproducible in large-scale synthesis or (in the case of novel natural materials) commercial cultivation? Will it not turn out to be too expensive to be useful? Will it be commercialized at all?

It is not surprising that few perfumers chose to engage in this effort and that most are far more willing to use the new materials marketed by reputable fragrance houses that come with at least a rudimentary set of directions and with the badge of established usefulness that is implied by their commercialization.

What is surprising is that the perfumers who are the explorers and discoverers rarely get the recognition they deserve, even within their profession. The survey of the contributions of new synthetic materials to the history of perfumery which was prepared by the Technical Commission of the Société des Parfumeurs de France* names, for each of some 50 materials cited, the chemists who first synthesized it. The perfumers who first incorporated these materials in a successful perfume are named in less than ten cases.

One instance provides an exception to this pattern of obscurity: the use by Edmond Roudnitska of Hedione in *Eau Sauvage*. This instance was remarkable not only for the success of Eau Sauvage itself and of Hedione which within 20 years became one of the most widely used of all perfumery materials, being used at levels exceeding 20% in some recent perfumes. It is remarkable because the odor of Hedione on the blotter, weak and vaguely like a jasmin tea, is so unexciting that a

*Comité Français du Parfum, *Classification des Parfums*, Paris 1990, pp. 36–43.

number of experienced perfumers (how many they were will never be known) who had a chance to evaluate methyl dihydro jasmonate before it was used in *Eau Sauvage* failed to recognize its potential.

There is a lesson here: The worth of a material is often not apparent upon examination on the blotter. This is particularly true of materials with low vapor pressure and with a low slope of the psychophysical function. These are the materials that appear weak on the blotter in their undiluted form but that may exhibit unsuspected powers of performance in compositions, upon the skin or on laundered fabrics. To improve the chances of recognizing the worth of such materials, one should examine all new aroma chemicals in the form of 10%, 1%, and occasionally even more dilute solutions, even if they do not appear particularly strong in their neat state.

There is yet another important principle exemplified by the use of Hedione in *Eau Sauvage*. The synthesis of methyl dihydro jasmonate occurred within the context of a study of jasmin, and Hedione, in its odor, represents an aspect of the jasmin complex. The idea to use it to make better jasmin bases is obvious. Its incorporation in a herbaceous citrus composition certainly was not. Creative breakthroughs come from nonobvious uses.

The importance of Hedione was in the possibilities it offered for creating novel odor effects. This is not always the case for important new aroma chemicals. Some are valuable because they permit familiar odor effects to be achieved at lower cost than was previously possible, with better stability in certain media, more favorable skin safety properties or for other "secondary" attributes, attributes other than odor itself.

One final observation in closing: We have remarked that it is perfectly human for perfumers to be more motivated to work with materials with an established record of production and provided with the badge of usefulness implied by commercialization than to experiment with laboratory samples of unexplored materials. Yet in companies that maintain research departments, such experimentation with unexplored materials—and the occasional discovery of the potentials for creative advance that may slumber in such materials—is one of the most valuable contributions a perfumer can make to an employer. The creative leading edge that is the main purpose of maintaining research departments within the fragrance industry can become reality only through an intensive and imaginative collaboration between the perfumer and the research chemist.

15

Constraints to Creation

Commercially used, perfume compositions become part of widely and freely sold consumer products. This fact imposes major constraints upon the perfumer's work. Above and beyond being aesthetically pleasing and technically sound, perfumes must not endanger the health of the people who use the products or who are engaged in their production; their detrimental effects upon the environment must be minimal; they must respect prevailing opinions about what is good or bad, right or wrong; and they must be in accord with the demand for value-for-money which is always alive at the level of the manufacturer of consumer products and ultimately with the consuming public.

Over time these constraints become part and parcel of the perfumer's work. They are embodied in the range of materials that the perfumer can use: they are tacitly implied in every briefing and naturally and universally respected by perfumers. To exemplify the point by a dramatic example: the extract from the skin of young women used by the hero in Patrick Suskind's novel *The Perfume* may be a wonderful perfumery material, but no perfumer considers him- or herself handicapped by the impossibility of using this extract.

However, in times when constraints are tightening, the new limitations have not yet achieved the status of a quasi law of nature, and perfumers, having to deal with them with deliberation, are much aware of being constrained in their creative work. Today we live in a period of distinct tightening of constraints in at least four areas. Public awareness of possible health hazards of trace materials including perfumes has dramatically increased; for the first time ever we are becoming cognizant of possible environmental effects of the manufacture, use, and disposal of products; the issue natural-versus-synthetic has acquired a new edge; and the economic pressures that come to bear upon the perfumers' daily work even in the world's rich countries are steadily increasing. In this chapter, we will consider the ways in which these changes affect creative perfumery.

HEALTH CONSIDERATIONS

Consumer-Related Constraints

The constraints that all responsible members of the worldwide fragrance industry observe today are spelled out in the guidelines issued by the International Fragrance Research Association (IFRA). These guidelines are regularly revised and updated as conditions demand. Because of their importance, they are summarized in the table below, along with the current versions of the Introduction and of a statement by IFRA regarding Application of the Guidelines.

INTRODUCTION (Version of December 1992)

These Guidelines deal exclusively with the use of substances and materials as fragrance ingredients. IFRA advises against the use of fragrance ingredients under conditions which may produce harmful effects. In pursuing this goal, the Technical Advisory Committee reviews scientific data produced by the Research Institute for Fragrance Materials—RIFM—as well as proprietary results and all other relevant data brought to its attention.

The recommendations of the Technical Advisory Committee of IFRA are based on data at present available. These recommendations will continually be updated as necessary when further data become available.

Recommendations for the quantitative restriction of ingredients are expressed in percentage of the fragrance compound. All ingredient restrictions are based on a use level of the fragrance compound of 20% in a consumer product. A fragrance compound formulated in this way is in accordance with the IFRA Guidelines; if a fragrance compound is to be used at more than 20% in a consumer product, the maximum limits of any restricted ingredient it contains must be lowered proportionally.

Unless otherwise stated in the ingredient recommendations, a fragrance compound which will be used at less than 20% in a consumer product may contain proportionally higher levels of restricted ingredients. In this case, fragrance suppliers should inform users that because of the presence of materials restricted by IFRA, this compound should be only used in appropriate concentration for well-defined applications. Such uses can then be considered in compliance with the IFRA Code of Practice. It is understood that the necessary information to be given to fragrance users does not include disclosure of fragrance formulas.

For non-skin contact consumer products (see footnote below) recommendations may allow higher limits for some restricted ingredients. In this case also, users should be informed that the fragrance compound should only be used in this specific type of products.

If combinations of phototoxic fragrance ingredients are used, the use levels have to be reduced accordingly. The sum of the concentrations of all phototoxic fragrance ingredients, expressed in % of their recommended maximum level, shall not exceed 100.

The fact that the Committee recommends not to use a certain fragrance ingredient does not exclude the use of a natural fragrance material, containing that ingredient, providing there are sufficient data supporting the safe use of that natural material.

Footnote on Non-skin Contact Products

For the purpose of complying with the IFRA Guidelines, the following consumer products are considered as non-skin contact products:

- Solid substrate air fresheners
- Plug-in air fresheners
- Membrane delivery air fresheners
- Insecticides (mosquito coil, paper and electrical)
- Toilet blocks
- Joss sticks and candles
- Plastic articles (excluding children's toys)
- Fuels

In contrast, the products given hereafter may involve skin contact and are excluded from the above:

- Household cleaning products
- Aerosols
- Detergents
- Shoe polishes
- Pot-pourri

- Carpet powders
- Liquid refills for air fresheners

APPLICATION OF THE GUIDELINES (Version of April 1989)

In order to apply effectively the Code of Practice of the International Fragrance Association (IFRA) to the manufacturing and handling of all fragrance materials, fragrance manufacturers should take measures so that all fragrance compounds offered for sale are in full compliance with the "Industry Guidelines to Restrict Ingredient Usage" as issued by IFRA. In cases where a subsequent change to these Guidelines changes the status of a fragrance compound, the manufacturers should so inform the purchasing party and provide all possible information in order to enable them to judge whether the fragrance compounds can be used in accordance with these Guidelines. If it does not conform to the Guidelines, then the manufacturer should furnish the purchasing party with an alternative fragrance compound prepared in accordance with these Guidelines.

IFRA Industry Guidelines to Restrict Ingredient Usage[1]

Substance	Maximal Recomm. Dosage[2]	Reason for Restriction
Acetylenic alcohols and their esters	0[3]	EEC directive
Acetyl ethyl tetramethyl tetralin (AETT)	0	neurotox.
5 Acetyl 1,1,2,3,3,6 hexamethylindane (Phantolid)	10%	phototox.
Acetyl isovaleryl	0	sensitiz.
Allantroot oil	0	sensitiz.
Allyl heptine carbonate	0.01%	sensitiz.
Allyl thiocyanate (mustard oil)	0	
Amyl cyclopentenone	0.5%	sensitiz.
Amyl vinyl carbinyl acetate	1.5%	sensitiz.
Angelica root oil	3.9%	photosensit.
Anisylidene acetone	0	sensitiz.
Asarone, *cis* and *trans*	0	biol. effects
Balsam Peru	0	sensitiz.
Balsam Peru oil, distillate	2%	sensitiz.
Benzene	10 ppm	carcinogen
Benzylidene acetone	0	sensitiz.
Bergamot oil	2%	phototox.
Bitterorange oil expressed	7%	phototox.

Substance	Maximal Recomm. Dosage[2]	Reason for Restriction
Butyl hydrocinnamic aldehyde, *p*-tert (Bourgeonal)	3%	sensitiz.
Butyl phenol, *para* tert	0	sensitiz.
Cade oil, crude	0	
Cassia oil	1%	sensitiz.
Chenopodium ambrosioides oil	0	EEC directive
Cinnamic alcohol	4%	sensitiz.
Cinnamon bark oil	1%	sensitiz.
Cinnamylidene acetone	0	sensitiz.
Citrus oils containing 5-methoxypsoralen	75 ppm[4]	phototox.
Collophony	0	sensitiz.
Costus root oil, absolute, concrete	0	sensitiz.
Cumin oil	2%	phototox.
Cyclamen alcohol	0	sensitiz.
Damascenones	0.1%	sensitiz.
Damascones	0.1%	sensitiz.
Diethyl citraconate	0	sensitiz.
Diethyl malate	0	sensitiz.
Dihydro coumarin	0	sensitiz.
6,7 Dihydrogeraniol	0	insuff. info.
Dihydro safrol	0	EEC directive
2,4 Dihydroxy 3 methyl benzaldehyde	0	sensitiz.
Dimethyl anthranilate	50%	phototox.
4,6 Dimethyl 8 *tert*-butyl coumarin	0	photosensit.
Diphenyl amine	0	sensitiz.
Ethyl acrylate	0	sensitiz.
Ethylene glycol monoethyl ether	0	sensitiz.
Ethylene glycol monomethyl ether	0	sensitiz.
Ethyl heptine carbonate	0.01%	sensitiz.
Fig leaf absolute	0	phototox.
Furfurylidene acetone	0	insuff. info.
Grapefruit oil expressed	20%	phototox.
Heptenal, *trans* 2	0	sensitiz.
Hexahydro coumarin	0	sensitiz.
Hexenal, *trans* 2	0.01%	sensitiz.
Hexenal diethyl acetal	0	sensitiz.
Hexenal dimethyl acetal	0	sensitiz.
Hydro abiethyl alcohol (Abitol)	0	sensitiz.
Hydroquinone monoethyl ether	0	depigment.
Hydroquinone monomethyl ether	0	depigment.
Hydroxy citronellal	5%	sensitiz.
Iso eugenol	1%	sensitiz.

Substance	Maximal Recomm. Dosage[2]	Reason for Restriction
6-Isopropyl 2 decalol	0	sensitiz.
Iso safrol	0	EC regulation
Juniperus sabina L. oil	0	EEC directive
Lemon oil, cold pressed	10%	phototox.
Lime oil, expressed	3.5%	phototox.
Marigold (Tagete) oil, absolute	0.25%	photosensit.
Menthadienyl formate	0.5%	sensitiz.
7 Methoxy coumarin	0	sensitiz. photosensit.
Methyl benzyl ketone	0	insuff. info.
6 Methyl coumarin	0	photosensit.
7 Methyl coumarin	0	photosensit.
Methyl crotonate	0	sensitiz.
p-Methyl dihydro cinnamic aldehyde	0	sensitiz.
4 Methyl 7 ethoxy coumarin	0	sensitiz.
Methyl heptadione	0.01%	sensitiz.
Methyl heptine carbonate	0.01%	sensitiz.
Methyl methacrylate	0	insuff. info.
Methyl *N*-methyl anthranilate (Dimethyl anthranilate)	50%	phototox.
3 Methyl 2(3) nonene nitrile	1%	sensitiz.
Methyl octine carbonate	0.01%	sensitiz.
Musk ambrette	0	phototox. neurotox. photosensit.
Mustard oil	0	
Non-2-inic acid esters[5]	0	insuff. info.
Oak moss absolute	3%	sensitiz.
Oct-2-inic acid esters[5]	0	insuff. info.
Orange oil, bitter, expressed	7%	phototox.
Pentylidene cyclohexanone	0	sensitiz.
Perilla aldehyde	0.5%	sensitiz.
Phantolid	10%	phototox.
Phenyl benzoate	0	sensitiz.
Propylidene phthalide	0.05%	sensitiz.
Pseudo ionone	0	sensitiz.
Pseudo methyl ionone	0	sensitiz.
Rue oil	3.9%	phototox.
Sade tree oil	0	acute tox.
Safrol[6]	0	EC regulation
Tagete (marigold) oil, absolute	0.25%	photosensit.

Substance	Maximal Recomm. Dosage[2]	Reason for Restriction
Tea absolute	0	sensitiz.
Tree moss extract, absolute, resinoid	3%	sensitiz.
Verbena absolute	1%	sensitiz.
Verbena oil	0	sensitiz.
Worm wood oil	0	

1. Status as of April, 1993.
2. Based on a use level of the fragrance compound of 20% in a consumer product.
3. Substance with max. rec. level 0 must not be used in fragrances at all.
4. Total bergapten (5-methoxy psoralene) level in the fragrance compound.
5. Except the listed esters with established limits.
6. Essential oils containing safrol are limited at 0.05%.

ADDITIONAL SPECIFIC IFRA GUIDELINES

A number of fragrance materials must be used only together with other materials:

Carvone oxide with spearmint oil (1:1)
Cinnamic aldehyde with eugenol (1:1) or *d*-limonene (1:1)
Cinnamic aldehyde–Methyl anthranilate Schiff base with eugenol (2:1)
Citral with *d*-limonene, mixed citrus terpenes or α-pinene (3:1)
Phenyl acetaldehyde with phenyl ethyl alcohol or dipropylene glycol (1:1)

The reason for these Guidelines lies in the "quenching" phenomenon: the sensitizing potential of these materials is absent in the presence of their companion materials. The proportions given are by weight and indicate the maximum proportion of the restricted material in the mixture.

IFRA provides specifications for purity regarding the following fragrance materials: Allyl esters, Cade oil rectified, Farnesol, Hexylidene cyclopentanone, Nootkatone, Sclareol, and Oils of the Pinacea species.

IFRA specifies admissible methods of production for the following fragrance materials: materials from opoponax and from storax, acetylated vetiver oil.

A number of finished goods houses have issued additional restrictions that supplement the IFRA guidelines. There are also a few countries, notably Japan, that have issued additional restrictions.

Worker-Related Constraints

To protect the safety of the operators who handle perfume oils in bulk, most countries have issued directives for labeling such oils. These usually include information regarding flammability and toxicity hazards. They do not normally affect perfumers in their task of creating perfumes.

ENVIRONMENTAL CONSIDERATIONS

Until quite recently the notion that the ecological impact of perfume oils presented an issue to be reckoned with was held by only a fringe group of environmentalists. Although perfumes are part of a great many products that are used daily by people in households throughout the world, quantitatively speaking they make up only a minute portion of most of these products. The worldwide annual consumption of aroma chemicals, including those used in the flavoring of foods, in 1990 has been estimated at 88,000 metric tons (SRI 1992). On a per-capita basis, this amounts to 0.05 g per day, which scarcely seems to warrant grave concern.

Moreover roughly 60% of this is accounted for by products that also occur in nature, in amounts that in most cases dwarf their use in synthetic fragrances, leaving only 0.02 g/person/day that could even potentially be problematic in the ecological sense. Fragrance certainly occupies a very low position in any listing of ecological hazards. However, fragrance by its very nature is part not only of the material, quantifiable world but also of the world of images and of the communication of benefits. From this perspective the demand that the fragrance satisfy whatever ecological demands are placed upon the product as a whole has its merits.

At time of this writing, it is impossible to estimate the future impact of environmental considerations upon the practice of perfumery. Many of the scientific questions raised, such as biodegradability and the reactions that perfume materials in the vapor phase may undergo or trigger in the atmosphere, still await exploration. We can do no more than outline the issues and their possible impact upon the perfumer's work.

The questions regarding the environmental impact of perfume materials may be broken down into those that concern their origin and production, those that concern their use, and those that have to do with their disposal.

Production of Perfume Materials from Renewable Resources

A key environmentalist demand is that we must, in the production of energy and raw materials, avoid wherever possible the use of resources that are not naturally renewable within a reasonable time span. Among the perfume materials, those that are directly obtained from plants by distillation or extraction obviously comply with this demand. Those that are synthesized starting from plant chemicals—be they simple derivatives such as vetiveryl acetate or the products of more complex reactions such as the many materials that can be synthesized starting from the components of turpentine—are also beyond reproach, although ecologists may insist that the reagents required for their synthesis (e.g., the acetic acid used in producing vetiveryl acetate) must also be obtained from renewable resources. However, a great many perfume materials today are produced starting from raw materials derived from the refining of crude oil and from natural gas. These do not fulfill the criterion of "renewability."

Should the demand for "renewability" become a common one in fragrance briefings, it would cause a switch, wherever possible, from oil- and natural gas-derived aroma chemicals to turpentine-derived ones. Demands that a high proportion of the components of perfumes be derived from renewable resources would not constitute a very severe constraint upon perfume creation, but demands that all or nearly all components be so derived would necessitate a radical reorientation including, say, the abandonment of all synthetic musks known today.

Environmental Impact of Aroma Chemical Manufacture

The demand that the actual and potential impact of chemical manufacturing plants upon the air, the groundwater, and the soil must be minimized already is law in the industrialized world. It does not directly affect perfumers in their work. It does, however, have the long-term effect of driving out most of the small- and medium-sized chemical manufacturers who cannot afford the steadily mounting environmental protection investment required by regulation.

Traditionally the several thousand aroma chemicals used by perfumers have included many for which aggregate world demand is on the order of a few tons per year or less. These were produced primarily by the small specialty manufacturers that are now disappearing. The range of aroma chemicals available for perfume creation is bound to shrink. Already many of the materials described in Arctander's standard book *Perfume and Flavor Chemicals* (1969) are no longer obtainable. Also the laws of competitive advantage indicate that the current concentration process within the aroma chemicals industry will favor a corresponding development among the customers of this industry, the fragrance houses—a development that is now in full swing.

Volatile Organic Chemicals

The discovery of the effect of fluoro chloro hydrocarbons upon the ozone layer of the atmosphere has had a profound impact not only upon the aerosol industry and the producers of propellants but also upon public and scientific thinking about the whole issue of atmospheric pollution. One of the consequences of this reorientation has been the introduction, in some American states, of regulations designed to control the use of all volatile organic chemicals (VOCs); these are organic chemicals that enter the atmosphere through evaporation. By their very nature all perfume materials fall into this class.

Already the VOC concern is prompting some manufacturers of household products such as air fresheners and window cleaners to change their formulations, thereby creating new challenges with respect to perfume solubility. It is impossible to estimate at this time to what extent in the future this concern will confront the perfumer with additional questions and challenges at the international level.

Biodegradability

The natural cycle of chemical substances involves a constant process of synthesis and breakdown of more or less complex structures. Living organisms, for example, ingest molecules of moderate complexity such as hydrocarbons and fats, and highly complex substances such as proteins. In metabolism they submit these to chemical transformations, which result partly in the synthesis of the molecular structures of which the organisms are made up (quite complex structures for the most part) and partly in the formation of extremely simple molecules such as carbon dioxide and water. After their death these organisms become themselves the substrate of the metabolism of other living organisms such as predators or microorganisms.

Before the dawn of synthetic chemistry the natural cycle had resulted in a fair degree of stability in the chemistry of the earth's surface, although in certain ecological niches, massive buildups of substances that would not break down (in those niches) have occurred, most notably in the deposits of mineral oil, natural gas, and carbon.

The development of chemical industry has provided us with the means to produce, on an ecologically significant scale, chemicals that interfere with the natural cycles of synthesis and breakdown either because they accelerate or slow down large-scale natural processes (e.g., the fluoro chloro hydrocarbons which accelerate the breakdown of ozone by sunlight) or, more commonly, because they resist breakdown themselves (e.g., certain synthetic polymers). This has become a matter of grave and widespread concern and has resulted in regulations and voluntary measures to restrict or prohibit the manufacture and use of materials that interfere with the natural cycles. This concern is particularly acute in those cases where this interference has direct or indirect adverse effects upon human health (as in the case of the fluoro chloro hydrocarbons), but it exists also where massive accumulation occurs without known health hazards (as in the case of the too-stable synthetic polymers).

Perfume materials have not been involved in any known cases of health hazards through accumulation in the environment. They are not a serious cause for concern regarding accumulation as such (without toxicity effects) because of the relatively small amounts in which they are produced and used. Nevertheless, for the reasons already discussed, the perfumer is sometimes faced with a demand for proposing fragrances with a high degree of biodegradability.

Natural microbiological degradation of complex materials down to water, carbon dioxide, inorganic salts, and bacterial "biomass" is a slow process that may require weeks, months, or even years for its completion. Its rate depends upon the conditions under which it takes place. To be meaningful, a specification for biodegradability must therefore be stated in terms of the proportion of the substance or mixture concerned that is broken down under specific conditions within a given time span. A widely accepted specification for "ready biodegradability" today states that 60% breakdown should occur within 28 days under the conditions of the "Closed-Bottle-Test OECD 301D" (OECD, no date).

As this book is being written, little public information is available regarding the biodegradability of individual perfume materials. This is due in part to a lack of agreement about appropriate specification. It is to be hoped that agreement will be reached on an internationally uniform basis and that, once this has been achieved, the testing of perfumery materials will be conducted as a joint effort by the fragrance industry in the same way as toxicity information is being gathered today.

In the absence of detailed information the following rule of thumb regarding biodegradability may serve to give general guidance:

Straight chain > Branched chain > Alicyclic > Aromatic

Straight chain compounds are most readily degraded, and aromatic compounds are most resistant to degradation.

CONSUMER CONCERNS

Natural Origin

Ever since 1828, when Friedrich Wöhler succeeded in synthesizing urea, chemists have held that the properties of materials, including their biological effects, were determined entirely by their chemical constitution and had nothing to do with their origin: All of the properties of a synthetic material are precisely the same as those of a naturally derived material with the same structure. The evidence for the correctness of this tenet that has been accumulated in the intervening 165 years is massive and irrefutable. Nevertheless, the feeling that natural materials are somehow fundamentally different from synthetic ones has never been dispelled among nonchemists.

Along with the growth of concern about technology going rampant and turning against humanity, the feeling has become widespread in the second half of our century that materials of natural origin are not only fundamentally different from synthetic ones, they are better: safer, more gentle, and richer. The marketers of consumer products have been quick to pick up this ground swell of popular feeling and have sought, wherever possible, to associate their products with nature rather than with chemistry. Many of the product categories that contain fragrance have also been affected by this trend. Shampoos have appeared on the market that are "free from harsh synthetic detergents," the claim "contains no artificial (or synthetic) colors" is widespread, and recently sun protection products "free from chemical sunscreens" have made their appearance.

In many cases marketers would also like to claim "free from synthetic perfume," but this desire is usually frustrated by the fact that for most popular odor types it simply is not possible to compose all-natural fragrances of good consumer acceptability at a reasonable price. What is possible is to create good perfumes of a wide range of odor types that contain a high level of natural components, say, more than 50% or 60%. If "nature identical" perfume materials, namely synthetic materials identical in their structure to naturally occurring ones, are also admitted, an even wider range, and more reasonable prices, become feasible.

From a practical point of view this is all that can and need be said about the subject. The remarks that are about to follow are of interest only to perfumers who would like to dig deeper, and to prepare themselves for discussions on the "synthesis-versus-nature" subject. We regret that they are full of provisos and apparent contradictions. In any debate this makes it very hard for the differentiated view to hold its own against the sweeping view, "nature is good, chemistry is bad."

While it is true that materials of identical *chemical structure* have the same odor qualities and are indistinguishable in their effect upon the skin irrespective of their origin, it is also true that very often synthetic and natural perfume materials that have the same *name* are not really identical in the chemical sense. Rhodinol from geranium oil consists of a different mixture of optical isomers than synthetic rhodinol. Eugenol from clove leaf oil is accompanied by different impurities than synthetic eugenol. Hence the natural and synthetic materials do differ in their odor quality, and they may differ in their effects upon the skin.

Although the essential oils, absolutes, resinoids, and so on, are universally regarded as natural perfume materials, we know today that their isolation from the plant tissue is accompanied by chemical changes, so that they are not really natural in the strict sense. The most gentle technique available for isolating essential oils from plant material is cold pressing; ironically, the two most widely used cold-pressed oils in perfumery, bergamot and lime, are not safe and must be treated before they can be used in perfumes.

Many dermatologists prefer synthetic perfume materials over natural ones. This is because the chemical constitution of the synthetics is totally known (provided they are free from impurities or we know what the impurities are), that of the naturals is not. Moreover, most naturals are mixtures of a great many different materials (e.g., more than 400 in the case of rose oil)—and the greater the number of materials in a mixture the greater the likelihood, statistically speaking, of having one or a few amongst them that can trigger an allergic reaction. In fact, among the perfume materials that have been restricted in their use, or eliminated altogether, for reasons of safety, there are a number of naturals: Balsam Peru, Oakmoss absolute, Costus oil, oil of Rue. On the other hand, there is the curious phenomenon of "quenching" whereby some materials that are irritating when used singly are no longer so when used together with other materials by which they are accompanied in the natural occurrences. For example, citral when used by itself has a greater potential for skin irritation than when it is used along with limonene. In lemon oil and other citrus oils, citral always occurs along with limonene.

Aided by chemical analysis, we have learned to produce close synthetic matches of many natural perfume materials; yet the last fine points still elude us, and we have never yet managed fully to reproduce the naturals, in all of their radiance and fullness. On the other hand, head space chromatography has enabled us to reproduce the odor of living flowers more faithfully than we ever could before; but these faithful reproductions of natural odors involves the use of materials that are not present in the "natural" essential oils or absolutes.

The progressive replacement of natural materials by their synthetic counterparts in contemporary perfumery is motivated in part by economic considerations, but there are other reasons as well:

- *Availability:* until recently, there was no known way to obtain, at any cost, the natural odor concentrates of muguet, lilac, gardenia and a great deal of other flowers; we do not know how to raise the whale population that would be needed to satisfy our demand for ambergris odor.
- *Creative freedom:* our modern perfumes contain rose and jasmin bases that smell like no rose or jasmin blossom has ever smelled, and the same goes for all other nature-inspired notes in perfumes. It is not just a matter of odor quality; we could never achieve the intensity and lasting power of today's perfumes using only naturals.
- Sometimes, we use synthetic materials out of regard for nature. The prime example are the synthetic musks which have saved the musk deer from extinction.

Fortunately, this rather confusing story has a simple moral: let us pay a great deal of attention to the odor effects of the materials we use, and let us worry a great deal about their safety upon the skin; but let us not give too much thought to their origin—and let us try to persuade our customers to see things the same way.

Animal Origin

Animal protection groups and consumers sympathetic to their cause are becoming increasingly vociferous in their demand that consumer products should not contain ingredients obtained by the killing or mistreating of animals. This concern, most widespread in the United Kingdom, applies also to cosmetic products and toiletries. It leads to a demand also for fragrance compounds free from materials of animal origin.

The requirement to avoid natural musk and ambergris places little or no strain upon current perfumery practice. Castoreum plays a minor role in perfumes for skin care products and toiletries and can easily be avoided. Whether the production of civet involves maltreatment of civet cats is a matter of debate. If it must be avoided, good synthetic substitutes are available.

PRICE

Value for Money

Today, looking back at 1953 home prices and apartment rents, hotel guides and restaurant menus, price lists for cars and for car repairs, we are inclined to smile wistfully. Until we remind ourselves of how much lower incomes were then, the era of the two dollar steak, the two hundred dollar city apartment rent, and the two thousand dollar car looks like a golden age indeed. Yet over the 40-year time span in which prices for most consumer goods and services have multiplied manifold, the raw material cost of an ounce of perfume and the cost of perfuming a kilo of detergent or any other personal or household product has remained essentially the same.

This has not been caused by a general lowering of quality. In fact, demands upon perfumes have definitely increased during this period, if not in terms of aesthetic quality then certainly in terms of human and environmental safety. Rather, the avoidance of cost increases has resulted from the conjunction of several technical and economic developments: lower raw material prices due to improved manufacturing processes and economies of scale, the lowering of import duties, more performing fragrances due to new aroma chemicals and the exacting demands of evaluation boards and clients, and sharper competition due to gas chromatography.* Taken together, these developments have resulted in an upward spiral of value for money that has continued unabated during the past four decades even if it cannot continue forever.

Four decades of essentially constant prices for fragrance compounds have created a climate of expectations on the part of the fragrance-

*A fuller explanation of this point is given in the Preface.

using industries that puts very high pressures upon the skills of the perfumer. Only the truly professional perfumer has a chance today of being successful over an extended period.

$/Odor Value Unit Compared to $/kg

The fragrance industry has always, and rightly, sought to convince its customers that in buying perfume compounds, not the cost per kilogram perfume compound counts but the cost of perfuming one kilogram or one ton of finished product. Much the same logic applies to perfumers when they examine raw materials to use in creating fragrances for an assignment. Not the cost per kilogram counts but the cost involved in achieving a certain odor effect in the finished product. Saccharin is far more costly than sugar on a per-kilogram basis, but in terms of achieving a specific degree of sweetness it is less expensive.

The best way of quantifying the performance of perfume materials in finished products available today is through the adjusted odor value units discussed in Chapter 13. Tables of such units are only now beginning to be constructed. Experienced perfumers, however, have tables of this kind, based upon extensive practical experience, in their heads. These mental tables may be imprecise and nonexplicit, they are nevertheless an essential tool in the perfumer's work.

AVAILABILITY

Availability of raw materials in the quality and quantity required is not usually a constraint facing perfumers. It may, however, become one under certain circumstances including the following:

- Experimentation that includes naturals from noncultivated plants.
- Projects of such magnitude that the amounts of certain components needed add up to a significant proportion of world production. When *Downey* fabric conditioner was launched, with a fragrance containing a fairly high proportion of patchouli, the world supply of patchouli was temporarily strained. With rare natural essences, shortages may arise due to far smaller projects.
- Projects that depend upon the manufacture of perfume oils in countries with import restrictions.

16

The Perfumer and the Market

The commercial success of a perfumer depends to a great extent upon an understanding of the market for which he or she creates perfumes. Perhaps we should say *markets* rather than *market* because most perfumers are called upon, in the course of their career and sometimes within the span of a single working day, to create perfumes for very different markets.

When we refer to markets in this context, we mean markets both in a geographic sense and in the sense of product categories. The shampoo market of Scandinavia is quite different in terms of odor preferences from that of Indonesia, and quite different from the Scandinavian household products market. Within Indonesia, there are actually two distinct soap markets, one dominated by international brands, and the other by traditional domestic products.

For the perfumer, a "market" is defined by a set of fragrance traditions, consumer expectations, and odor meanings. All of these are closely interrelated. Unless perfumers intimately know and intuitively understand the specific traditions and expectations and the "odor language" prevailing in the market for which they create, their chances for success are slim.

The first step in acquiring this knowledge and understanding consists in studying in depth the products that make up the market. The words "in depth" should not be taken lightly. Perfumers must study the minor brands as well as the major ones, the ones developed and produced by local manufacturers as well as the international ones, and develop an understanding of the differences that exist between these submarkets.

These may be differences of fragrance type but also differences of intensity or of tenacity. International brands may, for example, be characterized by sophisticated fragrances used at moderate concentrations resulting in medium strength and tenacity, while local traditional brands use simpler, less costly fragrances marked by great tenacity, at higher concentrations.

Perfumers must study the perfumes of the products that make up the market also in relation to their positioning. What fragrance types characterize baby soaps and distinguish them from family soaps? Do the perfumes of the different brands of medicated soap have something in common? What fragrance qualities set the premium priced products apart from the more common ones, heavy duty products from delicate ones? The answers to questions such as these define the odor meanings and the odor language that prevail in a market.

Perfumers must study these fragrances until they no longer know them just from the outside, as a more or less curious range of perfumes that correspond more or less closely to their personal preferences, but understand them from within, feeling why they are the way they are. Perfumers must speak the odor language of the market for which they are to develop perfumes not like a casual tourist but like a native or at least like a sympathetic, understanding visitor.

It is possible to arrive at such understanding simply by smelling products in one's studio. However, being physically present in the market, breathing its air and observing the way people live, helps a great deal. One understands the perfuming of laundry bars in the Philippines far better after having observed Filipino women doing their laundry in the nearby river or in a tub in the yard. To truly understand the differences in the perfuming of upper-class and popular products one must know the traditions and attitudes characterizing the relations between the classes.

A good written briefing may provide a great deal of information that is relevant to a project, but to do an outstanding job, perfumers must draw on an understanding of the market that goes well beyond the specifications of the briefing. The product or marketing manager who develops the briefing, living and working within the market, often does not spell out critical fragrance attributes because he or she takes them for granted.

This is particularly true for projects related to secondary brands. A brand introduced by a company that has the power rapidly to ensure wide distribution of a new product and the advertising budget required to establish it as a leading brand may be able to impose perfumes that do not correspond closely to established traditions, thus establishing a new frame of reference, a new "tradition." In fact, as a result of the "global brand" concept, this is happening in many markets today.

Less powerful brands depend for their acceptance upon an observance of established traditions. To the extent that the perfuming of such brands deviates from prevailing traditions, the deviation must "make sense" to the consumer within the positioning of the brand. Thus the first shampoo in a given market that adopts a "natural herb extracts" claim should reflect this differentiation in its perfume, but the perfume should be in line with how consumers in that market might expect natural herb extracts to smell.

SUCCESSFUL INNOVATION

Perfumers who know and understand the fragrances of the major brands in a market are in a position to make acceptable traditional perfumes for that market. Any perfumer who wants to deviate from tradition and create successful innovative perfumes must know his or her market so intimately as to have acquired an intuitive feeling for its consumers' olfactory frame of mind. Creating successful innovative perfumes for any market is the ultimate creative challenge.

People like what they know, and all innovation entails a risk. Part of the key to success lies in understanding how much innovation a specific assignment calls for, how much risk one can and should take. The Innovation Scale in Figure 16.1 gives some guidance with respect to this difficult question. It shows the cultural, product class related, and marketing plan related factors that jointly dictate how far one should move to the right, away from tradition, in any assignment. A few explanatory comments may be helpful.

Culture

Cultures differ a great deal in the respective value they assign to tradition and conventional wisdom, on the one hand, and to innovation and progress, on the other. These values are usually reflected also in the attitudes toward age and youth, in the respective powers of religion and science, and in the degree of rigidity of the social or class structure. The more the dominant values lie in the direction of progress, youth, science, and social mobility, then the more eager for or tolerant of innovation the culture will be also with regard to the products it uses and to their fragrance.

The Innovation Scale

Perfume	Traditional	Innovative within traditional	Radically innovative
Culture	Static		Dynamic or in transition
Product ambience	Ritual		Individual creative
Target group	Mass market		Selective
Advertising budget	Low		High
Examples	No-name products Local brands in developing countries	Major new brand introductions	Niche products Major fine fragrance introductions

FIGURE 16.1

In any given culture the openness to innovation may vary over time. It is particularly high in times of political upheaval, in the aftermath of a successful revolution or a lost war and during and after the periodic sociocultural paroxysms that we call "culture revolutions." The radically new life styles that swept Europe in the aftermath of the French Revolution, the adoption of chewing gum in Japan and of Coca Cola and hamburgers in Germany after World War II, and the manifold changes in many western markets after 1968 are cases in point. The instant success of the radically innovative "musk oil" which, in many guises, strongly influences western perfumery up to the present, would have been unthinkable without the coming of age of the post–World War II Flower Power generation in the 1960s.

Product Ambience

The context within which a perfumed product is used has a strong bearing on the degree of fragrance innovation that is tolerated or desired. In products used within a ritualistic context, tolerance of innovation is minimal. Experimental innovation in the scent of incense used at church services would be shocking and unthinkable. The same holds, to a but slightly less degree, for the fragrances surrounding the celebration of Christmas. In most cultures the family home has the emotional meaning of a safe haven and refuge from the turbulent and threatening world outisde. Nothing should ever change in the home.

Naturally this feeling limits the degree of innovation tolerated in household products.*

Sometimes, however, innovative fragrances may be appropriate even within the household product category. This is the case, for example, in products aimed at homemakers who see the home as a place in which to express their individuality, a place in which they give free reign to their creativity and imagination. Innovative fragrances may also be indicated in products used in conjunction with modern household devices. In a country where the population at large begins to purchase vacuum cleaners, new products to be used with the vacuum cleaner should probably have fragrances distinctly different from the cleaning product fragrances traditional to that country. When showers began to be widely installed in German homes in the 1960s and 1970s, this trend was accompanied by the rise of shower foams with fragrances distinct from both the traditional soap and the traditional foam bath types.

Target Group

Since people like what they know and the acceptance of unfamiliar things requires an effort, new products that aim at instant acceptance by large segments of the population must not, in their fragrance, depart far from the familiar. On the other hand, products aimed at people who consciously set themselves apart from the crowd should have nontraditional fragrances.

This exclusiveness is desirable the more the fragrance of the product is intended to be noticed not just by the user but also by others. In an expensive night cream, for example, the motives favoring a traditional fragrance (it connotes safety and purity) may well override the wish for setting it apart. However, in an Eau de Toilette, exclusivity, noncommonness, is likely to be the dominating motive. In fact the desire to set oneself apart is probably the most important driving force

*There are two additional reasons why tolerance for fragrance innovation in household products is usually limited. Fragrance often serves a signal that a certain job has been done. The lingering clean scent of a floor cleaner tells the visiting friend or mother-in-law and the husband returning home from the office that the floor has been cleaned. It can do so only if the floor cleaner fragrance is instantly recognized for what it is, in other words, if it is typical, traditional. Moreover the scent of products used in the kitchen must be compatible with a wide range of foods. This explains the popularity of lemon odors in kitchen cleaning products in general and in dishwashing products in particular in many cultures.

for innovation in alcoholic perfumery. As soon as a fragrance becomes too popular it becomes, for this very reason, unacceptable to the elite.

Advertising Budget

Advertising, in addition to drawing attention to a new product, also assures people that the product is good. Where advertising is lacking or sparse, the assurance that the product is good must come entirely from the product itself. Normally this assurance is achieved by making the product as similar as possible to other, well-known good products.

This mechanism operates powerfully in the realm of fragrance. Products with low advertising budgets and nonadvertised products usually mimic the established market leaders also in their fragrance. However, this is not always the case. When the back-to-nature movement became powerful in northern Europe in the 1970s and 1980s, for example, many nonadvertised products used the association with herbs and flowers rather than with market leaders in order to convince the consumer of their safety and goodness; this approach was reflected also in the fragrances selected for these products. Such developments are, however, rather exceptional.

New brand introductions supported by strong advertising campaigns, on the other hand, not only have the liberty to use more innovative fragrances, they are nearly forced to do so. Normally such products do not represent totally new product categories; they compete against established products and must convince users of these products to switch. To do so, they must present clearly perceptible points of superiority against the established products or at least points of difference. Very often fragrance provides such a point of difference, a function that becomes important, the smaller the perceptible differences in product performance are.

Since the marketing plans of mass-advertised products generally call for rapid acceptance for the product by broad segments of the consuming public, the fragrance should not, however, be too radically innovative. The appropriate range for fragrances for major mass-market product launches is therefore somewhere near the middle of the Innovation Scale, innovative without departing too far from the traditions established by current leading mass-market brands.

The Interaction of Countervailing Tendencies

In establishing the most appropriate range of innovation, perfumers must consider all factors, those that relate to the culture, to the product ambience, the target group as well as the marketing plan. Often not all of these will point in the same direction; the appropriate fragrance should then lie somewhere between the extremes of a totally traditional fragrance and radical innovation. The case of major mass-market launches, where the need for clear differentiation is counterbalanced by the need for rapid widespread acceptance, is but one example among many. To understand the different factors involved, and to assign the appropriate weight to each, requires mature marketing judgment.

PERFUMER, EVALUATION BOARD, CLIENT, AND PUBLIC

Perfumers may perfectly understand the attitudes and expectations of the market for which they create their fragrances; they may succeed perfectly in developing fragrances that meet the expectations and yet fail—because their creations never reach the market. This can happen because a perfumer does not offer his or her perfume, or the product for which the perfume was developed to the public. Between the perfumer and the public there stand at least two groups of people who must be convinced that the fragrance is the right one before it has a chance to reach the market. One of these groups is within the perfumer's own company, the other is within the client company that markets the product. Both groups may vary widely in their composition.

Most major perfume suppliers today have within their organizations a group of specialists who evaluate the submissions of the different perfumers who have worked on a project, often drawing upon off-the-shelf formulations in addition to specially developed ones, and decide which fragrances should be submitted to the client. Depending upon the organizational structure and the situation, others within the perfumer's company may also have a voice in this decision: the chief perfumer, the application laboratory responsible for stability testing, sales personnel, or top management.

The group of people at the client company who are involved in the decision on which of the several (often very many) fragrances submitted for a product launch or relaunch should be finally adopted is even more varied. It may include the purchasing department, development chemists, evaluation panels attached to the development laboratories, marketing, in-house evaluation panels, the market researchers who design and interpret consumer tests, even top management—or, as the rumor often goes, the president's wife or secretary.

For the sake of this discussion, let us call the group involved at the perfumer's company the "evaluation board" and the totality of all of

The Gatekeeper Model

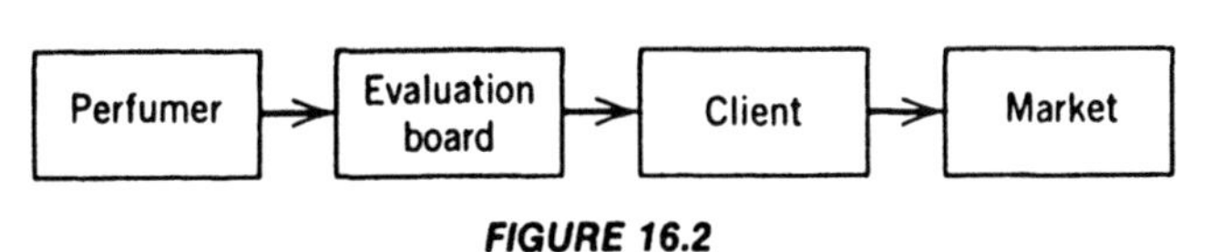

FIGURE 16.2

the people involved at the client level, the "client." We then arrive at the model shown in Figure 16.2. It is a typical example of what in current marketing literature is referred to as a "gatekeeper model." Unless there is good agreement among the perfumer and the various gatekeepers about what the market wants (Figure 16.3*A*), the perfume will not reach the market (Figures 16.3*B* and 16.3*C*). If there is good agreement among the perfumer and the gatekeepers (e.g., if both evaluation and the perfumer understand precisely what the client wants) and all of them are wrong, the perfume or the perfumed product may reach the market, but it will fail (Figure 16.3*D*).

Achieving the situation shown in Figure 16.3*A* requires high skills of understanding and of communication among all the parties involved, as well as organizational structures that favor open and effective communication. Moreover the longer the chain of gatekeepers—it may in actuality entail several successive gates both within the perfumer's and the client's companies—the lower are the odds for the arrow, the perfumer's creation, to strike home.

One link in the decision chain where a poor alignment of "gates" has been rather common in the past has been the consumer tests conducted by market research departments. Since fragrances, especially the fragrances in functional products, exert their influence upon the consumer at a subconscious level, special and sophisticated marketing research techniques are required to obtain a correct picture of this influence. Moreover the techniques of marketing research have been developed largely within the context of mass-market marketing and must be radically adapted to be appropriate to selectively marketed products (Jellinek 1991b).

The perfumer, alas, is usually confined to a rather passive role in these processes, with limited power to affect the decision mechanisms in his or her own company, and usually none at all to affect them at the customer level. The perfumer can only do his or her utmost to understand the criteria that guide the selection process at both levels and to take them into account in the perfume's creation.

Occasionally, however, a perfumer or a spokesperson close to him or her in the organization (e.g., the perfumery manager or an experienced sales executive) succeeds in making the perfumer's views heard

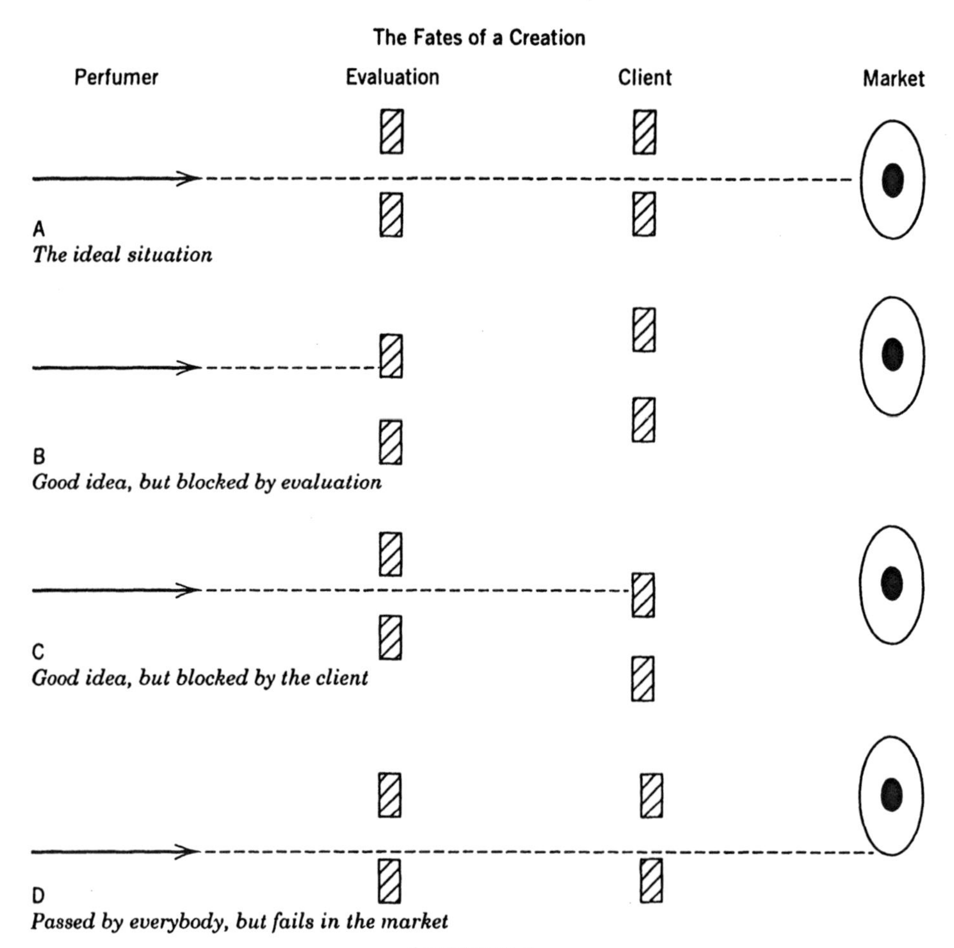

FIGURE 16.3

and respected at all stages of the decision process. Then the perfumer can exert a major influence upon the final decision. To attain this status requires force of personality, excellent communication skills, and a good track record. A classic example of this kind was Ernest Shiftan, chief perfumer of Van Ameringen-Haebler, later IFF, who developed a relationship of trust with Estée Lauder and exerted a major influence upon the highly successful selection of fragrances launched by that house for over a quarter century, from Youth Dew in 1952 until his death in 1977. The high degree of innovation that marked these fragrances was possible only because of the presence at the client level of a strong and courageous decision maker, Estée Lauder herself.

Part V

Scientific Fundamentals

17

The Chemical Structure of Perfumery Materials

The odor, volatility, strength, and stability of the materials used in perfumery are all determined by their chemical structure. It is useful therefore, and of interest, for the perfumer to have some knowledge of the various types of structures that occur, and of their naming and classification. For example, what is meant by the words "phenyl," "ethyl," and "alcohol," and how do these relate to the odor and other characteristics of the material known as "phenylethyl alcohol"?

In perfumery we are concerned almost entirely with **organic** materials, that is to say, with compounds made up of the elements **carbon** and **hydrogen**, usually also with **oxygen**, sometimes with **nitrogen**, and rarely also with sulphur or chlorine. Compounds, or molecules, based on atoms of carbon and hydrogen only are described as **hydrocarbons**. To this category belong many of the so-called **terpenes**, such as pinene and limonene, found in essential oils. To those materials that contain oxygen belong such groups as the **alcohols, aldehydes, ketones, lactones, acids**, and **esters**, while those containing nitrogen include the **nitromusks, anthranilates, nitriles, quinolines**, and **indol**.

BONDING AND CHEMICAL STRUCTURE

The chemistry of carbon is based on the fact that the carbon atom is **tetravalent**—that is to say, it always forms four **bonds** when combining with other atoms to form molecules. Hydrogen forms one bond, oxygen two, and nitrogen three. For example, the following combinations can occur:

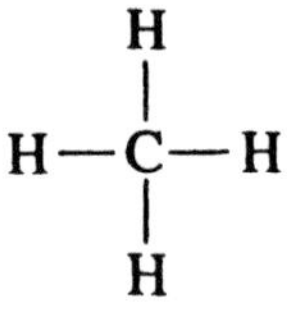

```
     H              H              H    H
     |              |              |   /
  H—C—H         H—C—OH         H—C—N
     |              |              |   \
     H              H              H    H
```

Methane 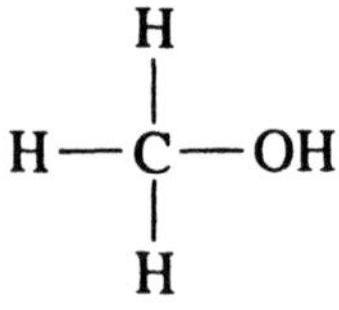   Methyl alcohol 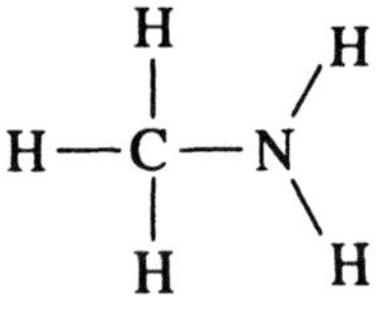   Methylamine

Carbon atoms can also bond to each other to form extended **chains** of atoms linked to each other. These chains may be either **straight** or **branched**. For example,

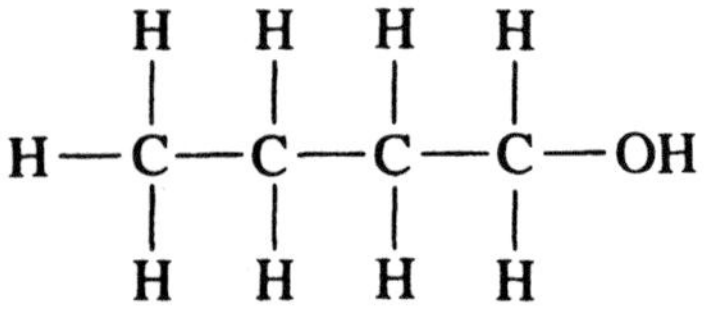

```
     H   H   H   H
     |   |   |   |
  H—C—C—C—C—OH
     |   |   |   |
     H   H   H   H
```

written $CH_3(CH_2)_2CH_2OH$

n-Butyl alcohol

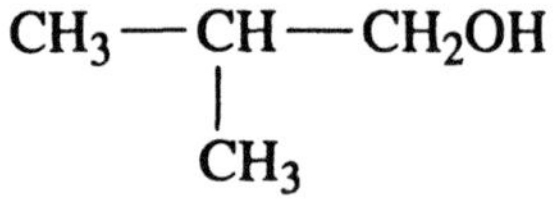

```
CH3—CH—CH2OH
     |
     CH3
```

or $(CH_3)_2CHCH_2OH$

Isobutyl alcohol

Note that the two forms of butyl alcohol are made up of the same number and types of atom, C, H, and O, arranged in different ways.

It is also possible for two of the potential bonds to be shared between a pair of atoms, between two carbon atoms, or between a carbon atom and an oxygen atom. For example,

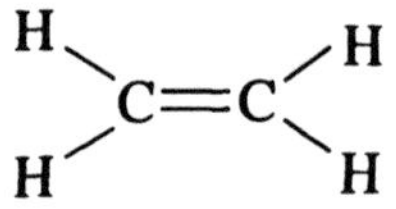

```
 H       H            H
  \     /             |     O
   C=C            H—C—C//
  /     \             |     \
 H       H            H      H
```

Ethylene 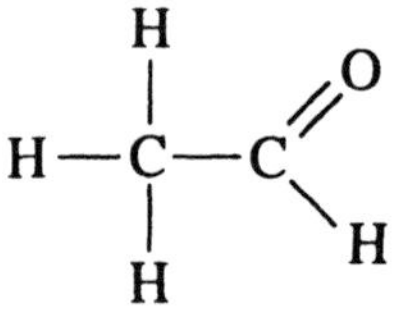   Acetaldehyde

These are called **double bonds**, and a molecule containing a double bond between two carbon atoms is said to be **unsaturated**. By the addition of two hydrogen atoms across a double bond, **hydrogenation**, the molecule is said to be **saturated**. For example,

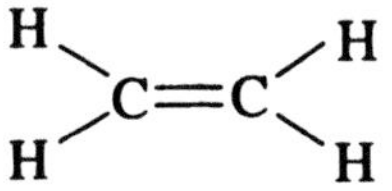

$$H_2C{=}CH_2 \qquad\qquad H_3C{-}CH_3$$

Ethylene—unsaturated 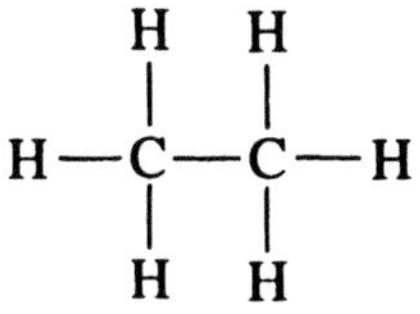   Ethane—saturated

Similarly 2 nonenal (iris aldehyde) and 2,6 nonadienal are both unsaturated forms of aldehyde C9, the aldehyde with nine carbon atoms in the chain:

$$CH_3(CH_2)_7CHO \qquad\qquad CH_3(CH_2)_5CH{=}CHCHO$$

Nonanal (aldehyde C9)   2 Nonenal

$$\underset{9}{CH_3}\underset{8}{CH_2}\underset{7}{CH}{=}\underset{6}{CH}\underset{5}{CH_2}\underset{4}{CH_2}\underset{3}{CH}{=}\underset{2}{CH}\underset{1}{CHO}$$

2,6 Nonadienal

Note that the suffixes *-anal*, *-enal*, and *-dienal* signify the saturated form and the forms with one and two double bonds, respectively; note also that the numbering of the carbon atoms begins from the —CHO, or aldehydic group, at the end of the chain, with the double bonds occurring at the 2 and 6 positions.

Triple bonds may also occur between carbon atoms. For example,

$$HC{\equiv}CH \qquad\qquad CH_3{-}O{-}\overset{\displaystyle O}{\overset{\|}{C}}{-}C{\equiv}C(CH_2)_4CH_3$$

Acetylene   Methyl heptine carbonate

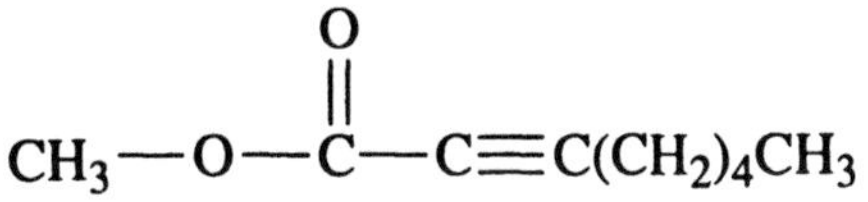

or between a carbon atom and a nitrogen atom. This is characteristic of the **nitriles**. For example,

$$CH_3(CH_2)_5CH{=}CHC{\equiv}N$$

Iris nitrile

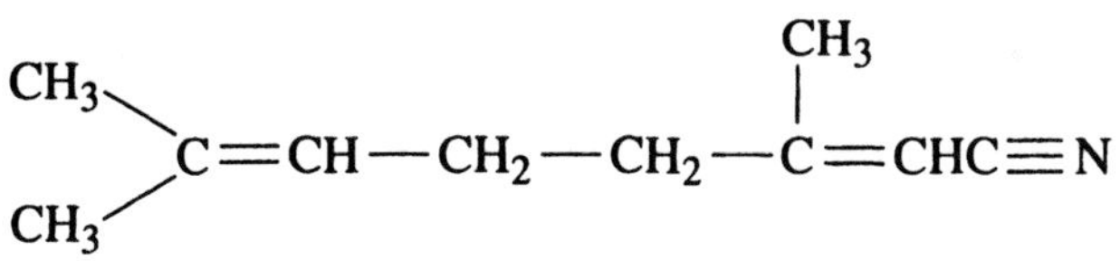

$$(CH_3)_2C{=}CH{-}CH_2{-}CH_2{-}\overset{\displaystyle CH_3}{\overset{|}{C}}{=}CHC{\equiv}N$$

Geranyl nitrile

Carbon atoms are also capable of joining together to form **ring structures**, the most common of which is the six-membered **cyclohexyl** ring. This may occur in the saturated form, or with a double bond between one or two pairs of carbon atoms in the ring, or with three alternate double bonds loosely shared between all six carbon atoms. It is usual, when writing out the structures of ring compounds, to use a form of chemical shorthand. Some perfumery examples are given below:

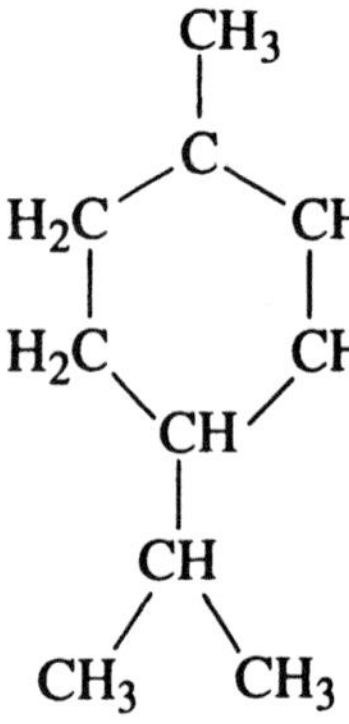

Menthol

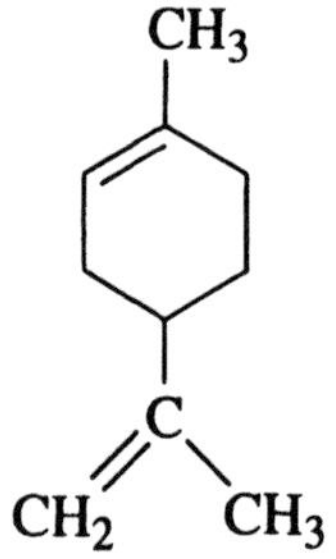

Limonene

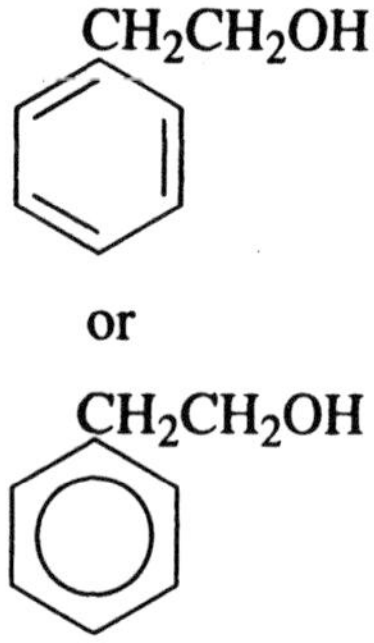

Phenylethyl alcohol

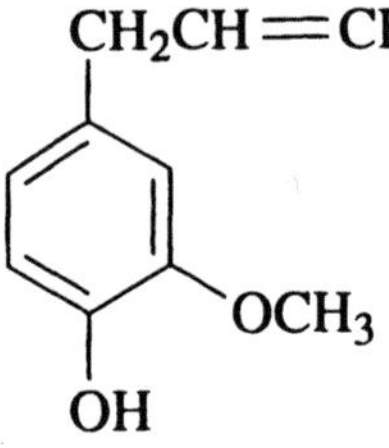

Eugenol

This last type of ring, with three double bonds, known as the **benzene ring**, is of great importance in perfumery, since it occurs in a wide range of materials, both natural and synthetic. Materials whose structure is based on the benzene ring are described, for historical reasons, as **aromatic**—not to be confused with the same word when used either in its original sense of having an "aroma," or as a perfumery description for a particular type of smell, associated with "aromatic" herbs.

Ring structures of five atoms may occur, or even up to sixteen or more, in which one member of the ring is sometimes oxygen. Large ring structures of this type, such as in cyclohexadecanolide, are described as **macrocyclic**. Similarly nitrogen may occur as part of a ring structure, or two or more rings may be linked together:

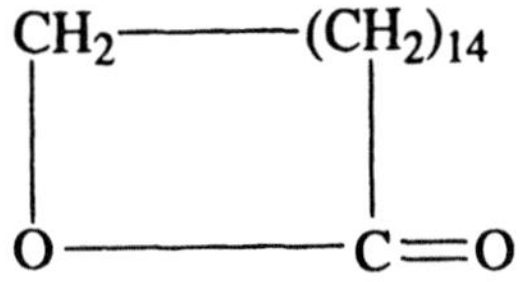

CH_2——$(CH_2)_{14}$ | O——C=O

Cyclohexadecanolide

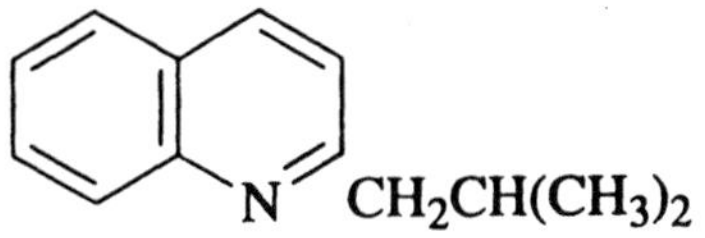

N $CH_2CH(CH_3)_2$

Iso-butyl quinoline

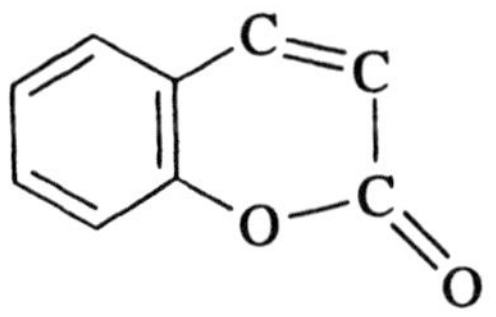

C=C | O—C=O

Coumarin

It will be appreciated from this very brief look at the various types of linkage possible between the atoms of carbon, hydrogen, oxygen, and nitrogen, that the number of combinations possible is almost limitless. Indeed more than eight million organic compounds are already known to exist. The whole of life depends on such diversity—not just perfumery.

A list of the most important structural groups found in perfumery materials is given in Appendix A.

The Terpenes and Their Derivatives

The terpenes form one of the most important groups of perfumery materials, including such products as geraniol, linalool, terpineol, camphor, cedrene, and their many derivatives. They occur widely in nature, and some are used as starting points for the synthesis of other materials, such as the ionones, and many of the modern woody chemicals. Before their structure and chemistry was fully understood, the terpenes were defined simply as the insoluble constituents of essential oils. It was found that the majority contained either 10 or 15 carbon atoms, the two groups being named **mono-** and **sesquiterpenes**. Other terpenes containing higher multiples of five carbon atoms were also known to exist in both plants and animals. The structure of these materials was subsequently explained by the working out of their biosynthesis.

All terpenes are in fact formed by the linking together of two or more units of five carbon atoms, originally thought to be molecules of isoprene, or **isoprene units**. This description is still used, although it is now known that the chains are based on the combination of two slightly different materials, isopentenyl pyrophosphate and dimethylallyl pyrophosphate, which combine to form geranyl pyrophosphate. The addition of a further isopentenyl pyrophosphate molecule produces farnesyl pyrophosphate. These two materials, which are directly related to **geraniol** and **farnesol**, form the starting points from which all other mono and sesquiterpenes are derived:

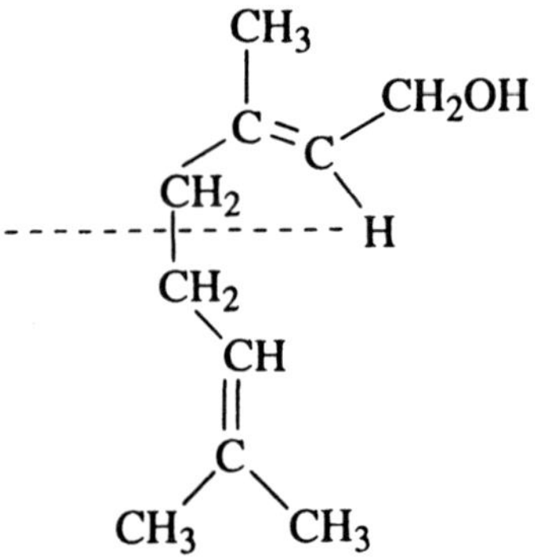

$$(CH_3)_2C{=}CH{-}CH_2{-}CH_2{-}C(CH_3){=}C(H){-}CH_2OH$$

Geraniol

$$\begin{matrix} CH_3 \\ CH_3 \end{matrix}\!\!>C{=}CHCH_2 \,\vdots\, CH_2\underset{\substack{|\\CH_3}}{C}{=}CHCH_2 \,\vdots\, CH_2\underset{\substack{|\\CH_3}}{C}{=}CHCH_2OH$$

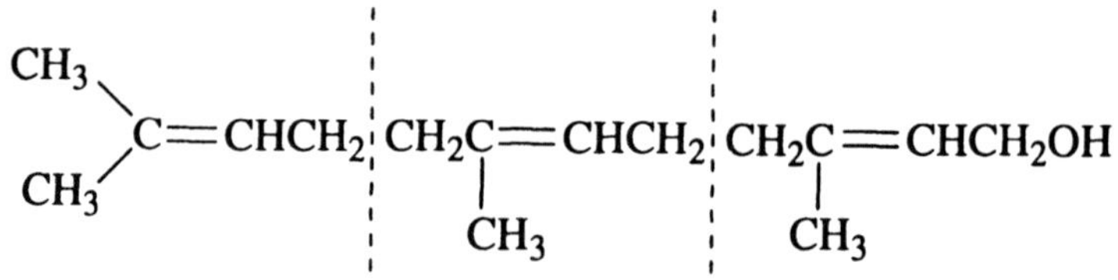

Farnesol

Although, chemically, all products belonging to these two series are correctly described as terpenes, in perfumery (just to make matters even more complicated!) the word is often used in a narrower sense to describe only the hydrocarbon members of the series. These include limonene and terpinolene with 10 carbon atoms, and cedrene and caryophyllene with 15. Materials belonging to the series but containing oxygen are then described as **terpenic alcohols, aldehydes**, and so on.

Many of the terpenes, such as terpineol, limonene, and cedrene, are ring structures formed by the rearrangement of straight chain molecules:

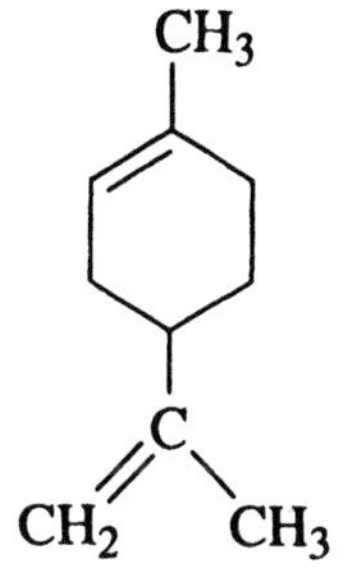

Limonene

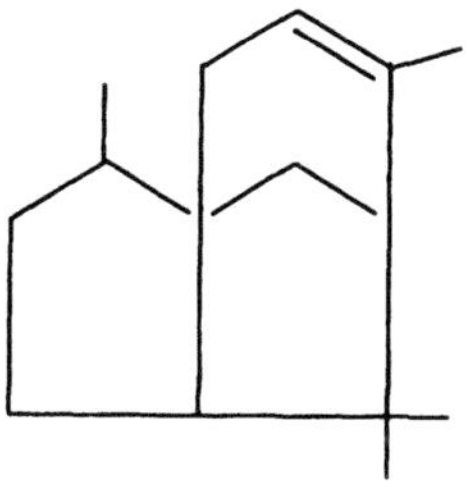

Cedrene

Isomerism

Two or more compounds made up of the same number and types of atoms but with different chemical structures are known as **isomers.** We have already seen the difference between the two isomers *n*-butyl and isobutyl alcohol. These are known as **skeletal isomers**, since they have different carbon skeletons. There are also **positional isomers** in which the "functional group" is located at a different position on the carbon skeleton. For example, alpha, beta, and gamma terpineol all have the same arrangement of carbon atoms but with the —OH (hydroxy) group and the double bond in different positions:

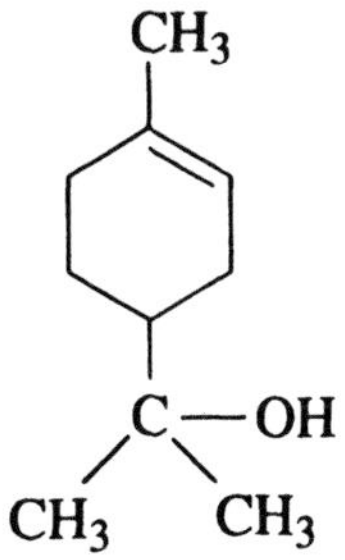

alpha-Terpineol

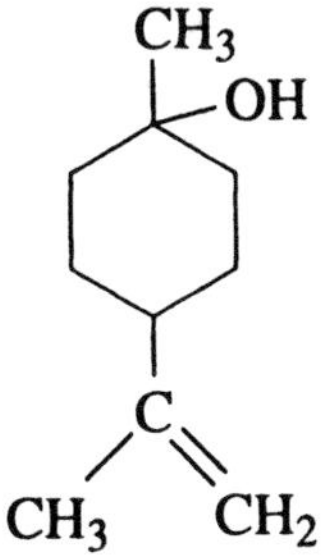

beta-Terpineol

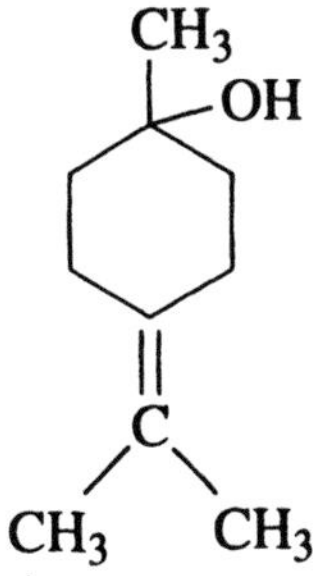

gamma-Terpineol

In citronellol the difference between the naturally occurring alpha form and the synthetically produced beta form lies only in the position of one double bond:

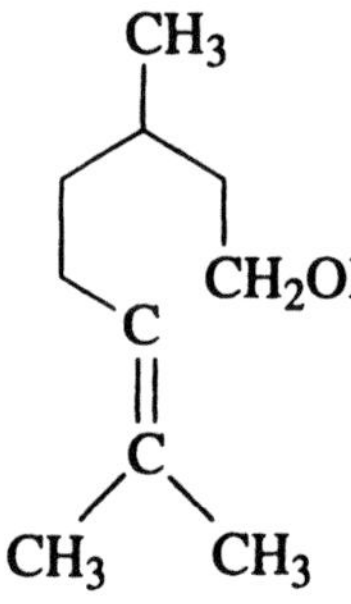

alpha-Citronellol

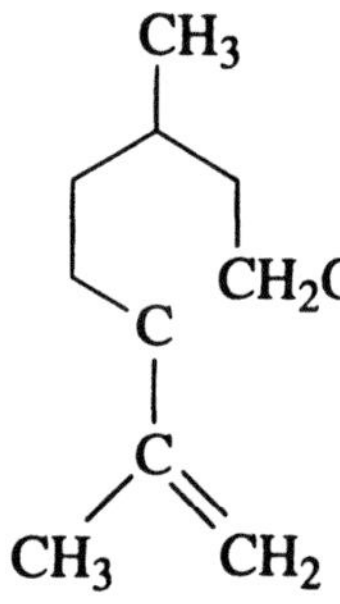

beta-Citronellol

Another type of isomerism is that of **stereoisomerism**. This is caused by the arrangement of atoms around a double bond; all the connections are the same, but there is a difference in the three-dimensional orientation. When a double bond occurs between two carbon atoms, the two atoms are not free to rotate about their common axis. For example, two forms exist for the following hypothetical molecule:

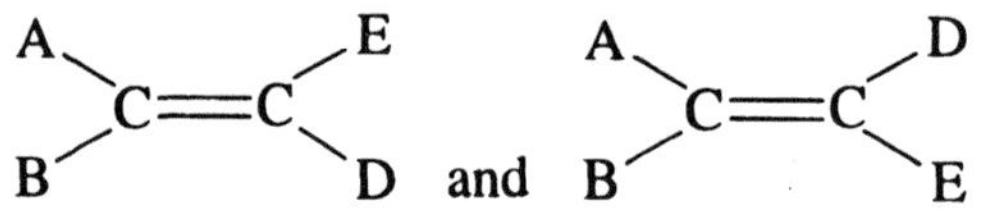

$$\mathrm{(A)(B)C{=}C(E)(D)} \quad \text{and} \quad \mathrm{(A)(B)C{=}C(D)(E)}$$

This form of isomerism occurs between geraniol and nerol which are otherwise identical:

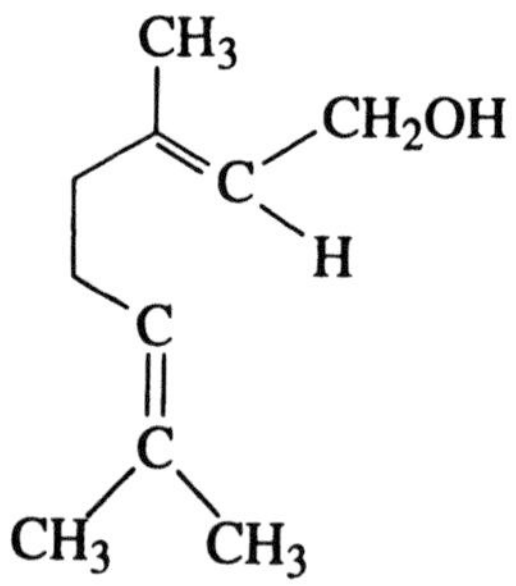

Geraniol

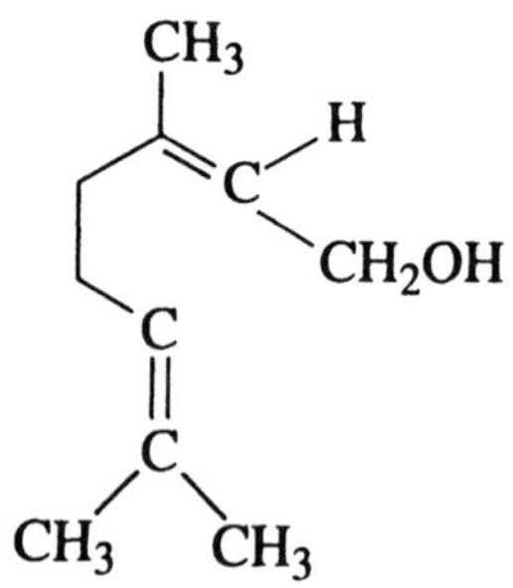

Nerol

So-called ***cis-trans*** materials are also examples of stereoisomerism, the prefix *cis-* coming from the Latin meaning "on the same side" and *trans-* meaning "across." For example,

$$\underset{\text{cis-3-Hexenol}}{\overset{H\qquad\qquad H}{\underset{CH_3CH_2\qquad CH_2CH_2OH}{C{=}C}}}\qquad\qquad \underset{\text{trans-3-Hexenol}}{\overset{H\qquad\qquad CH_2CH_2OH}{\underset{CH_3CH_2\qquad H}{C{=}C}}}$$

cis-3-Hexenol

trans-3-Hexenol

Cis-forms are more common in nature than *trans*-, and they tend to be more interesting in odor. They are, however, less stable chemically due to the uneven stress placed on the double bond by both larger groups being on the same side of the molecule.

Finally, a mention must be made of **optical isomerism**. It is observed that some materials can exist in two forms: one that rotates the plane of polarized light to the right and one that rotates it to the left. These are known as the **dextro** and **levo** forms. The explanation of this phenomenon lies in the fact that the molecules can exist either in a right-handed or a left-handed configuration. This is a difficult concept to visualize but may best be illustrated by comparing the two hypothetical compounds represented three-dimensionally below.

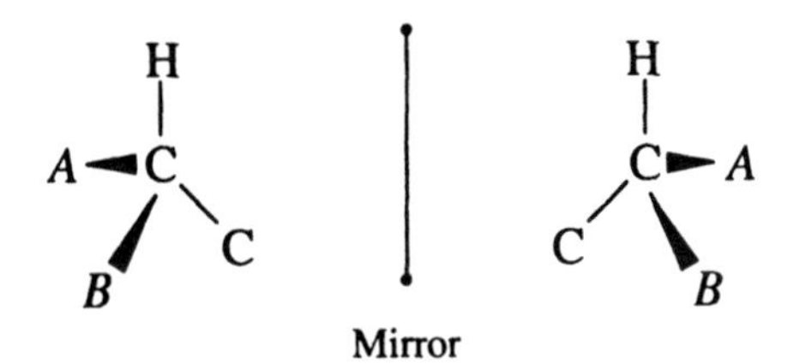

The two molecules are identical in every way except that one is the mirror image of the other. When two such isomers occur in equal proportions, the mixture, which has a net zero optical rotation, is said to be **optically inactive**, or **racemic**.

Often the difference in odor between two such forms is surprisingly great. For example, dextro carvone is the main odor constituent of caraway oil, while its optical isomer levo carvone is typical of spearmint. Among other materials that occur in important dextro and levo forms are citronellol and rose oxide.

FUNCTIONAL GROUPS AND THEIR RELATION TO ODOR

A functional group may be defined as a group of atoms, part of a larger molecule, whose bonds have a characteristic chemical behavior.

In whatever molecule it occurs the group will behave in approximately the same way. The **classification of chemical substances** is based on the functional groups that they contain. Examples of such groups are double bonds, the aldehydes, alcohols, and esters. A list of the more important of these, together with their nomenclature and chemical formulas, is given in Appendix A.

From the point of view of the perfumer this classification can be very instructive, since it provides not only information as to the probable stability of materials in various types of products but also assists in the understanding of their odor characteristics. Although the odor of a chemical substance appears to be determined by its overall shape and structure, the presence of a functional group both modifies the odor and brings to it its own particular character. The degree to which the functional group influences it depends upon the size, structure, and odor characteristic of the molecule, and the position of the functional group in it. This is best explained by considering a few examples.

1. Aldehydes and Alcohols

Aldehydes are, in general, stronger than their corresponding alcohols, as for example is aldehyde C12 lauric when compared to its corresponding alcohol. However, the degree to which these differences are expressed varies considerably from one aldehyde-alcohol pair to another.

Example 1(a) In the case of **benzyl alcohol** and **benzaldehyde** the molecular size is comparatively small with only 7 carbon atoms. The alcohol is also a weak material having virtually no odor. Here the change brought about by replacing the alcohol group with the aldehyde is quite dramatic.

Moving up the structural series to **phenylethyl alcohol**, which has one more carbon atom and a positive though not very strong odor, the corresponding aldehyde, **phenylacetaldehyde**, although very much more aggressive, and "green," in odor still retains some of the rosy character of the alcohol. In the case of **phenylpropyl alcohol** (hydrocinnamic alcohol), which has 9 carbon atoms and a very much more powerful odor, the difference between this and phenylpropyl aldehyde is more one of comparative strength than of odor type:

CH_2OH

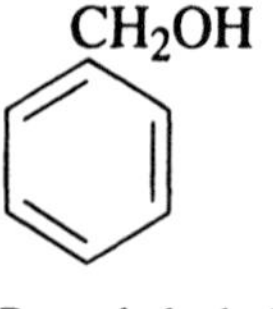

Benzyl alcohol

CH_2CH_2OH

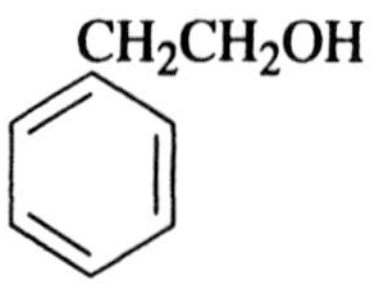

Phenylethyl alcohol

$CH_2CH_2CH_2OH$

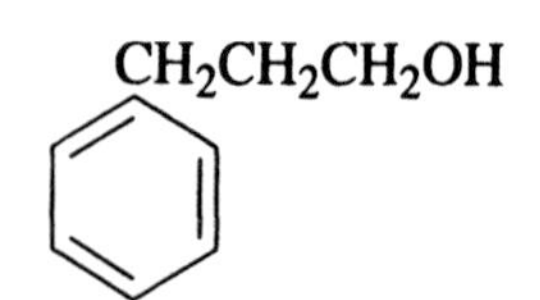

Phenylpropyl alcohol

Higher still up the series are **amyl cinnamic alcohol** and **aldehyde.** These are comparatively large molecules containing 14 carbon atoms, and their odors are therefore relatively similar in type, although there is still a very considerable difference in strength. Here the characteristic aldehydic note is almost missing from the aldehyde, though this may be due partly to the protective influence of the side chain. In **hexyl cinnamic aldehyde**, which has one more carbon atom in the side chain, the aldehydic note is virtually missing altogether. (The molecule is also chemically more stable.)

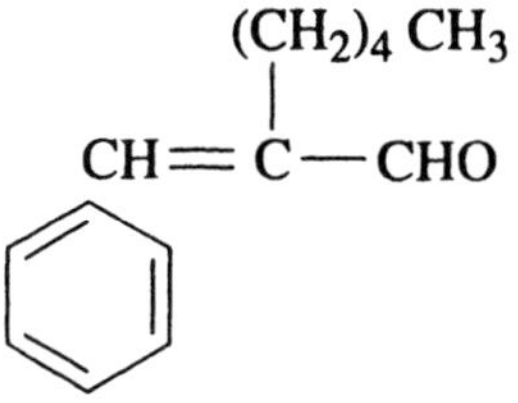

$(CH_2)_4CH_3$
CH=C—CHO

Amyl cinnamic aldehyde

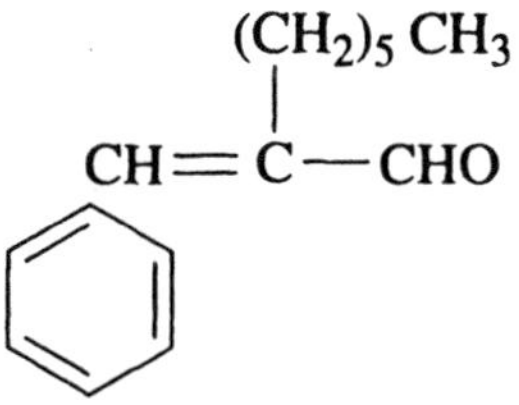

$(CH_2)_5CH_3$
CH=C—CHO

Hexyl cinnamic aldehyde

Example 1(b) The characteristic influence of aldehydic and alcoholic groups is also clearly shown in the relationship between **citronellol, citronellal**, and **hydroxycitronellal**.

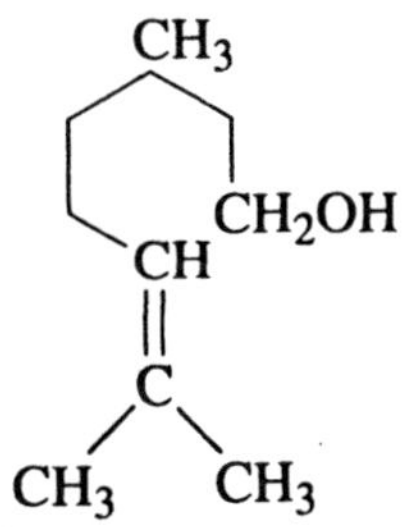

CH_3
CH_2OH
CH
C
CH_3 CH_3

Citronellol

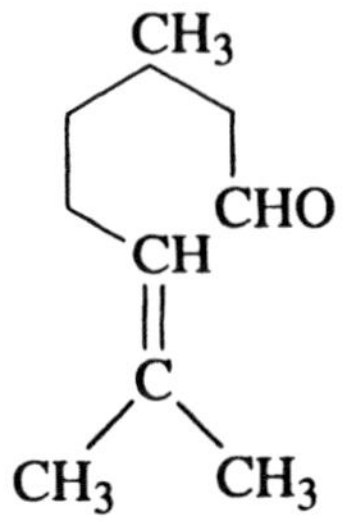

CH_3
CHO
CH
C
CH_3 CH_3

Citronellal

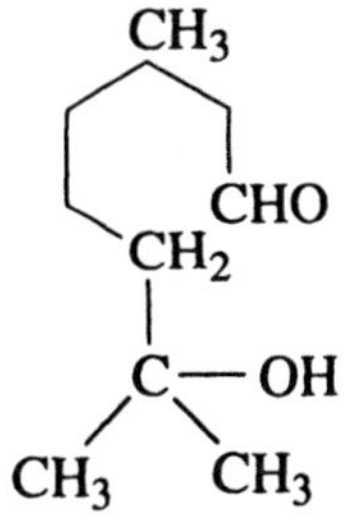

CH_3
CHO
CH_2
C—OH
CH_3 CH_3

Hydroxycitronellal

The difference between citronellol and citronellal is typical of that between an alcohol and an aldehyde. Although they are clearly related in odor, the aldehyde is, as we would expect from quite a small molecular structure, very much more powerful and harsh. In hydroxycitronellal the addition of an —OH and —H across the double bond (in effect by adding a molecule of water, a process known as **hydration**) produces an alcohol group near to the other end of the molecule, away from the aldehydic group. The material still maintains some of its aldehydic character but now, in addition, has some of the softness and floral character associated with an alcohol. The presence of more than one oxygen atom in the molecule also has the effect of greatly reducing its volatility.

2. Esters

The esters form a particularly interesting group in that each molecule is composed of two structural units joined together by the characteristic ester formation. One of the two component structures comes from an **alcohol** and the other from an **acid**. For example,

$$\underset{\text{Ethyl alcohol}}{C_2H_5OH} + \underset{\text{Acetic acid}}{HO-\overset{\overset{\displaystyle O}{\|}}{C}-CH_3} \longrightarrow \underset{\text{Ethyl acetate}}{C_2H_5O\overset{\overset{\displaystyle O}{\|}}{C}-CH_3} + \underset{\text{Water}}{H_2O}$$

In effect a molecule of water is formed as the two molecules combine—a type of reaction known as **condensation**. The resultant odor is a combination of the typical fruity-floral tendency of the ester grouping together with the individual characteristic of each of the two structural units. Dominance between the three parts of the molecule depends upon their relative size and odor strength.

How this applies in each individual case makes a fascinating study. A few representative samples may be given.

Example 2(a) **Ethyl acetate**, which is a comparatively small molecule, has the typical fruity character associated with all the lower esters, and a more or less equal balance between the influence of the two structural units, derived from ethyl alcohol and acetic acid. **Linalyl acetate** and **geranyl acetate**, however, although retaining the typical character of an acetate have much less of the ester fruitiness and are more closely related to the corresponding alcohols, linalool, and geraniol. The dominance of the alcohol appears to be even greater in **phenylethyl acetate** and **paracresyl acetate** (a phenolic ester).

Example 2(b) In the reverse case of the ethyl esters, where the alcohol derived radical is comparatively small, it is the acidic part of the molecule that dominates, as in the higher members of the series such as the **benzoate, cinnamate, phenylacetate**, and **salicylate**. Many of these higher ester radicals are very dominant in character and give rise to closely related series. Cinnamates and phenylacetates, for example, are some of the most readily recognized perfumery materials owing to their very characteristic odors.

Example 2(c) Esters derived from the weakly smelling **benzyl alcohol** are much less closely related to each other than those derived from more powerful alcohols such as linalool, geraniol, paracresol, or even phenylethyl alcohol. The higher esters of benzyl alcohol, in particular, seem to owe little to the character of the alcohol.

Example 2(d) In esters where both parts of the molecule are relatively large, the typical fruitiness is almost entirely absent, as for example in **benzyl salicylate, phenylethyl phenylacetate**, and **linalyl cinnamate**:

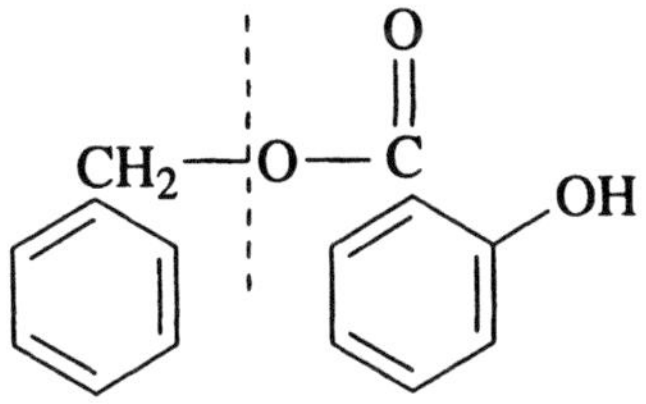

CH_2—O—C(=O) ... OH

Benzyl salicylate

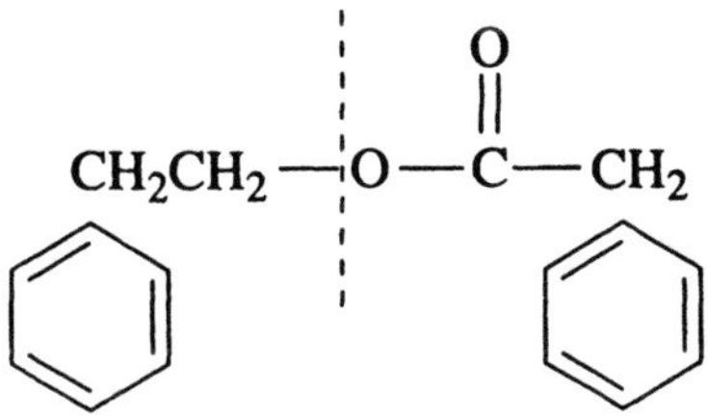

CH_2CH_2—O—C(=O)—CH_2

Phenylethyl phenylacetate

O—C(=O)—CH=CH

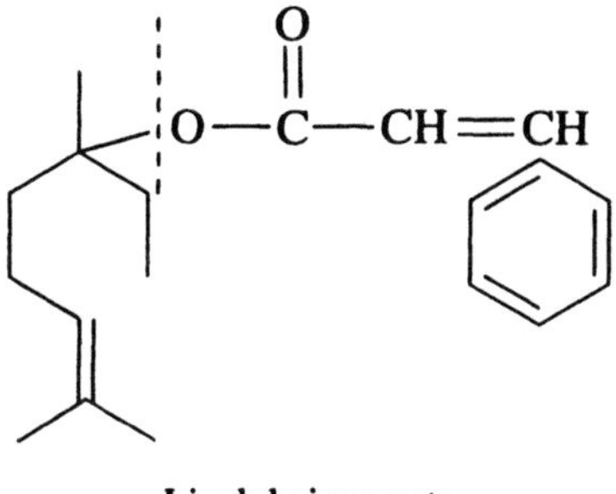

Linalyl cinnamate

3. Double Bonds

The presence of a double bond can have a significant influence on the strength and odor of the overall molecule, as well as in modifying the effect of another functional group.

Example 3(a) The dominance of the double bond in the ***cis*-3-hexenyl** series of esters, as well as in the alcohol (leaf alcohol) and aldehyde, results in a closely related group of materials in which the other functional groups play only a modifying role in the odor, despite the small size of the C6 chain:

$CH_3CH_2CH{=}CHCH_2CH_2OH$

cis-3-Hexenol

$CH_3CH_2CH{=}CHCH_2CH_2OOCCH_3$

cis-3-Hexenyl acetate

Example 3(b) The effect of a double bond on the odor of a molecule depends very largely upon its position. The three straight chain **aldehydes** with 11 carbon atoms provide a good example of this. The formulas are as follows:

$$CH_3(CH_2)_9CHO$$

Aldehyde Cll undecylic

$$CH_2{=}CH(CH_2)_8CHO$$

Aldehyde Cll undecylenic

$$CH_3CH_2CH{=}CH(CH_2)_6CHO$$

Intreleven aldehyde

The introduction of the double bond into **undecylic** aldehyde to give the **undecylenic** form produces a significant change both of character as well as of strength. If the double bond is now moved to a position between the third and fourth carbon atoms from the end of the chain (the C8 position counting from the functional group), giving **intreleven** aldehyde, the odor remains very similar but the strength increases enormously. The strength of a double bond at this position relative to the end of the chain is quite characteristic. As has already been seen, it occurs also in ***cis*-3-hexenal** (at C3), and in **2,6 nonadienal** (at C6), two unsaturated straight chain aldehydes of great olfactory strength. The strength of ***cis*-jasmone** may also be attributed partly to the presence of a double bond in this position.

Example 3(c) When two double bonds lie on either side of a single bond, the double bonds are said to be **conjugated**. This can occur in a number of different types of molecule, giving rise to such groups as conjugated aldehydes, or the conjugated triple bond of a nitrile:

$$=C-C=C \qquad C=C-C(H)=O \qquad C=C-C\equiv N$$

Double bond — Aldehyde — Nitrile

In perfumery this can have an important effect both on the odor of a material and its functional properties. Conjugated aldehydes tend to be sweeter and deeper in olfactory character than their saturated counterpart, for example, as with **citral** and **citronellal**, the aldehydes related to geraniol and citronellol, and **cinnamic aldehyde** and **phenylpropyl aldehyde** (hydrocinnamic aldehyde).

In the cinnamic molecule there is an additional conjugational effect between the double bond in the side chain and the unsaturated benzene ring. The saturated form, phenylpropyl alcohol is again the harsher of the two in character.

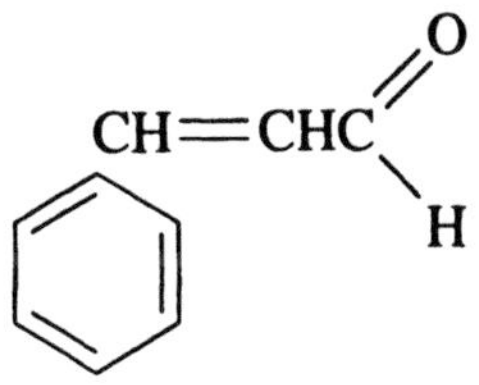

$C_6H_5CH{=}CHCHO$

Cinnamic aldehyde

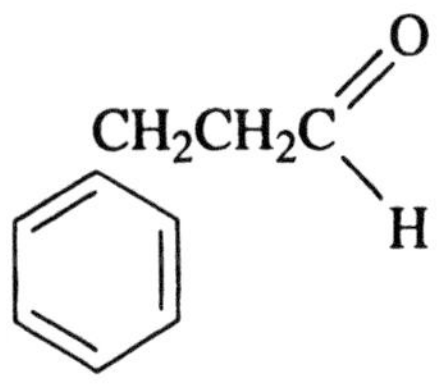

$C_6H_5CH_2CH_2CHO$

Phenyl propyl aldehyde

$C_6H_5CH{=}CHC(=O){-}CH_3$

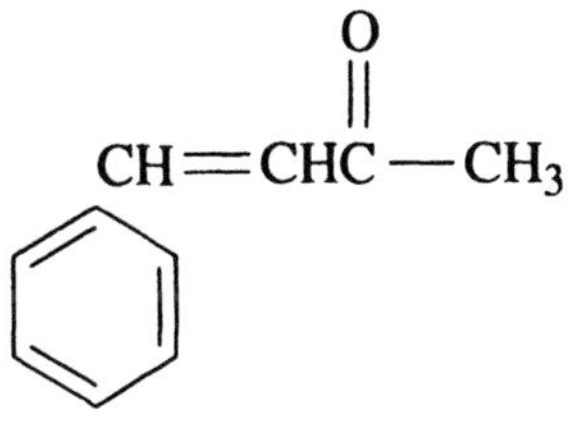

Benzylidene acetone

The conjugation of a double bond with a **carbonyl group**, C=O, such as in an aldehyde, can have the serious adverse effect of greatly increasing the likelihood of the material being a severe **skin irritant** or **sensitizer**. Examples of this type of influence occur in **citral, cinnamic aldehyde**, and ***trans*-2 hexenal**, as well as in **methyl heptine carbonate** which has a conjugated triple bond. **Benzylidene acetone**, which is closely related in structure to cinnamic aldehyde, can no longer be used as a perfumery material for this reason, and the level at which other potential sensitizers can be used in perfume formulations is restricted (see pp. 183–185).

18

Chemical Reactions in Perfumery

BENEFICIAL REACTIONS

Perfume compounds are essentially mixtures of materials that blend together rather than react. There are, however, a few important exceptions, some of which are used deliberately by perfumers as part of their technique in formulation, while others may be regarded as part of the normal process of maturation.

Schiff Bases

A Schiff base is formed by the combination of an aldehyde and an amine ($—NH_2$); usually in perfumery with methyl anthranilate. This is a condensation reaction in which water is produced. For example, **aurantiol**, the most widely used of the Schiff bases, is formed by the condensation of **hydroxycitronellal** and **methyl anthranilate**:

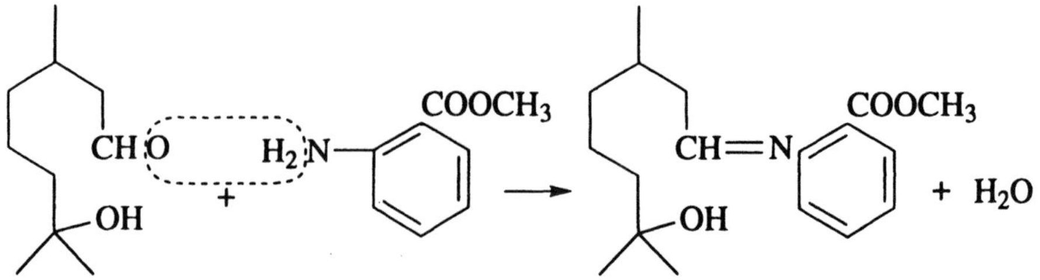

CHO + H_2N— (COOCH$_3$) → CH=N (COOCH$_3$) + H_2O

Hydroxycitronellal | Methyl anthranilate | Aurantiol | Water

Other important Schiff bases include those derived from **citronellal, Lyral, Helional**, and **Canthoxal**. One of the more recent and interesting is **Meaverte**, the product of **Tripal** and **methyl anthranilate**. Like most other Schiff bases these products are all strongly yellow in color—a useful way in which to recognize them in a compound.

Although Schiff bases are widely used as raw materials in formulation, they are also frequently allowed to form within the compound by the use of methyl anthranilate as a straight ingredient in combination with aldehydes and ketones. Schiff bases formed in this way often give a better result than is obtained by including them in the formula. The process is, however, a gradual one, and it may take several weeks or months for the Schiff bases and the resulting color to develop—a serious problem for quality control! Many of the modern **tuberose perfumes** are created using this technique, with both methyl anthranilate and ready-made Schiff bases being used at the same time with aldehydes such as hydroxycitronellal, Lyral, Helional, and Lilial. The final result after maturation is indeed a complex mixture.

Dilution has a significant effect in reducing the rate of the reaction, and the formation of Schiff bases is slowed down, although not entirely stopped once the compound has been diluted in alcohol. Methyl anthranilate occurs naturally in many essential oils together with aldehydes such as citral without forming Schiff bases owing to the low concentration of both materials.

Hemiacetals

Aldehydes and **ketones** react with **alcohols** to form loosely joined molecules known as **hemiacetals**. This is a reversible reaction that, depending upon the conditions, reaches a point of equilibrium in which both the free aldehyde and alcohol occur together with the hemiacetal. This is a somewhat complicated reaction, with intermediary products being formed, but it may be expressed simply as follows:

$$>C{=}O \quad -\overset{|}{\underset{|}{C}}-OH \quad \rightleftharpoons \quad >\overset{OH}{\overset{|}{C}}-O-\overset{}{C}<$$

Aldehyde/ketone — Alcohol — Hemiacetal

Some of the harsher aldehydes such as phenylacetaldehyde are quite often kept diluted in phenylethyl alcohol so as to take advantage of the softening effect of the hemiacetal formation before incorporating them into a compound.

Much of the initial maturing, or "getting together," of a perfume compound depends upon this formation of hemiacetals, and when diluted in ethyl alcohol, a further period is required for the formation of hemiacetals between the ingredients of the compound and the base. In the production of fine fragrances this process is usually allowed to continue for several weeks prior to chilling and filtering, with the length of time having an important effect upon both the quality and persistence of the end product.

HARMFUL REACTIONS

Unfortunately for perfumers, most of the reactions with which they are concerned tend to be destructive to their compounds rather than beneficial. Perfumes are complicated mixtures of materials, all of which are liable to change under certain conditions and to react among themselves as well as with the products into which they are put. Exposure to air, heat, and light can be damaging to their stability, as can the presence of metals such as iron. Perfumes often have to be designed for use in hostile bases that may be either strongly acid or alkali. Perfumers concerned with such products will spend much of their time assessing the stability of individual materials, so as to eliminate those that are unstable, which often prove to be the majority. Even this type of test may prove to be inconclusive, since the base may act as a catalyst causing the perfumery materials to react among themselves. At very low perfume levels, however, this effect is unlikely to be great.

The whole subject of perfume stability is an enormous and complex one, and it is only possible here to give some indication of the various types of problem that perfumers are likely to encounter and to suggest a few of the possible solutions where problems exist. We may begin by looking at the stability of two of the most widely used groups of perfumery materials, the aldehydes and ketones.

Aldehyde and Ketone Stability

Aldehydes and ketones are comparatively reactive materials and can be the cause of many stability problems in perfume compounds. Two of the most important types of reaction that are specific to these materials are the formation of acetals with alcohols, and the so-called aldol reaction in which two molecules of the same aldehyde or ketone combine with each other to form a larger one.

The Formation of Acetals Unfortunately, the beneficial formation of hemiacetals, which has already been discussed, can under acidic conditions be carried a stage further leading to the formation of acetals:

Hemiacetal Acetal

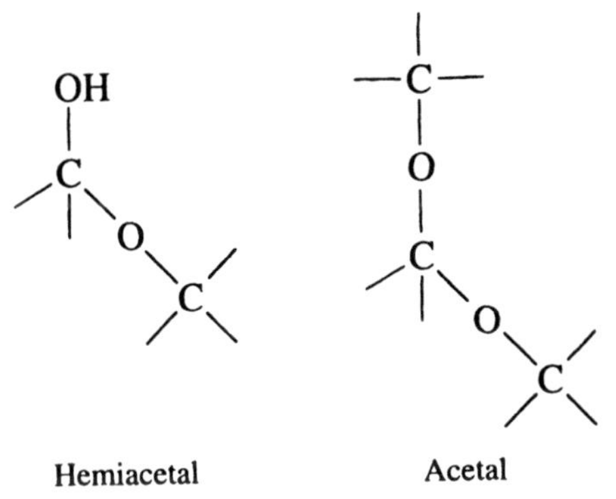

As we will see later, the release of acids is characteristic of the breakdown of esters in the presence of water. This acidity, in alcoholic fragrances, can in turn result in the formation of acetals between the ethanol and any aldehydes that may be present in the perfume. Although this reaction is reversible in principle, where there is a large excess of alcohol in the product as opposed to aldehydes, as is usually the case, the aldehydes can be almost entirely lost, with a catastrophic effect on the odor.

When examining an alcoholic fragrance by gas chromatography, it is quite usual, if an old sample has been used, to find large amounts of these acetals, such as hydroxycitronellal di-ethyl acetal and the di-ethyl acetals of aliphatic aldehydes, which were never meant to be there in the original formula and which would not have been present if a fresh sample had been used. The presence of large amounts of glycols in an acid medium can similarly lead to the formation of diglycol acetals. These are often virtually odorless compared with the original aldehydes, and their formation can result in the total loss of what was intended to be the most powerful part of the fragrance.

The Aldol Reaction It will probably have been noticed by anyone who has worked in a perfumery laboratory that **aldehyde C12 lauric** and **phenylacetaldehyde** tend to solidify with age. What is happening is that the molecules are combining to form larger molecules that make up solids rather than liquids. This reaction can be shown in the following formula for phenylacetaldehyde:

$$2\ C_6H_5CH_2CHO \longrightarrow C_6H_5CH_2CH(OH)-CH(C_6H_5)CHO$$

Phenylacetaldehyde

This type of reaction, known as the **aldol reaction**, which occurs particularly in the presence of alkali, is reversible, resulting in an equilibrium between the proportions of single and combined molecules.

Not all aldehydes behave in precisely the same way under these conditions depending upon the one of three types to which they belong. These may be represented as follows:

$$\begin{array}{c} H \\ | \\ C-C-CHO \\ | \\ H \end{array} \qquad \begin{array}{c} H \\ | \\ C-C-CHO \\ | \\ C \end{array} \qquad \begin{array}{c} C \\ | \\ C-C-CHO \\ | \\ C \end{array} \text{ and } \begin{array}{c} \\ \\ C{=}C-CHO \\ | \\ C \end{array}$$

Type 1   Type 2   Type 3

Chemically these are described as mono-, di-, and tri-substituted acetaldehydes.

In the first type, which includes **aldehyde C12 lauric** and **phenylacetaldehyde**, the balance of the reaction lies in favor of the combined molecule. In the second, however, to which belong **aldehyde C12 MNA, cyclamen aldehyde, hydratropic aldehyde**, and **Lilial**, the balance lies more toward the original material. Most ketones also behave in this way. The third type, which includes **amyl** and **hexyl cinnamic aldehydes**, do not undergo the aldol reaction and are therefore more stable. Aldehydes such as **anisaldehyde** and **Triplal**, in which the —CHO is linked directly to a ring structure, are also less reactive.

This, however, is not the end of the story. Under slightly more extreme conditions, or over a period of time, the products of the aldol reaction can go on to form other more complex materials. As a result the aldehyde may be completely lost. Some of the products formed in this way are strongly colored and can therefore cause discoloration in the end product.

Such reactions can have a considerable effect on the stability of aldehydes and ketones in bases that are even mildly alkaline, such as soaps and detergents. Before soap making reached the standards of quality that are achieved today, the base would often contain, and still does in poorer qualities, a surplus of alkali from the process of saponification. This made it almost impossible to use any of the less stable types of aldehyde with any degree of confidence. Now, however, with good quality bases being available most aldehydes, including many of the straight chain aliphatic aldehydes, can be used quite safely.

A similar type of reaction occurs between aldehydes, ketones, as well as lactones, and **ethyl acetoacetate**. This material is sometimes used in industrial deodorants and masking agents precisely because of its ability to react with the unpleasant smelling aldehydes, and other reactive materials, that they are designed to cover. When ethyl acetoacetate is included in a perfume formula for this type of product, often at quite a high percentage, together with aldehydes, the shelf life of the compound tends to be somewhat limited owing to this reaction. Ethyl acetoacetate and a few related materials are also used in fine perfumery, though usually in such small amounts that the effect of dilution reduces the risk of any serious reaction.

Ester Stability

Esters, as has already been seen, are formed by the combination of alcohols and acids with the elimination of water. The reverse reaction, which is one of **hydrolysis**, can also occur under certain conditions, such as the presence of acid or alkali in an aqueous base. Not all esters are equally susceptible to this type of breakdown, but it only requires the formation of traces of a free acid such as butyric acid, totally to spoil a compound. Even an acetate such as linalyl acetate can cause problems by giving rise to free acetic acid. In alkaline media progressive breakdown of the ester may occur as the acid becomes neutralized. In products where this is likely to be a problem terpinyl acetate is often used to replace linalyl acetate.

Oxidation

Most perfume compounds containing essential oils will deteriorate when exposed to the air, and a number of different reactions can be involved, all of which come under the general heading of oxidation, brought about by the presence of oxygen. Many of the unsaturated monoterpenes, which occur widely in citrus, coniferous, and seed oils, are especially liable to oxidation, forming in the first place peroxides by the addition of oxygen across the double bonds, which can then break down to give a variety of different products.

Compounds that are in direct contact with air when in the final product, for example as in powder products or in soap, are particularly at risk from oxidation, and the head space above the product is likely therefore to become rich in the breakdown products of oxidation. For this reason the packaging usually allows for a certain amount of "breathing" so as to let these "off-odors" escape rather than their being reabsorbed into the product. A number of products exist, generally classed as **antioxidants**, for example, BHT (butyl hydroxytoluene), that help to inhibit these oxidation reactions. These antioxidants are often added to citrus oils, or to compounds to prolong their shelf life, or to the final product.

Another method used to prevent the possibility of oxidation is to flush out the air from the container in which the material is to be stored with an inert gas such as nitrogen. This method is most frequently used for large containers such as drums, particularly if these are only partly full.

Metal Contamination

The presence of certain metals such as iron, particularly in the presence of water, can act as a **catalyst** for many types of chemical reaction including oxidation, and iron can itself form highly colored reddish-brown complexes with many materials. A common test for the presence of iron in essential oils such as patchouli, where it can occur as a result of the distillation process being carried out in nonstainless steel equipment, is to mix the oil with benzyl salicylate. If iron is present, a reddish color develops after a few moments. Fortunately iron can quite easily be removed by treatment with citric, tartaric, or oxalic acids, with which it forms stable sequestered complexes. Compounds that show this type of discoloration can sometimes be recovered in this way if not too old, whereas products containing water (which may be the source of the iron contamination), such as alcoholic lotions, will usually show too much odor deterioration, by the time the discoloration has developed, to make recovery possible.

The Effect of Light

Light is a source of energy that is needed for certain types of chemical reaction. Some of these so-called **photochemical reactions** are of value to the perfumery industry, being used for the synthesis of such materials as the **rose oxides**. However, a number of perfumery materials discolor badly in the presence of ultraviolet light. Some of the nitromusks, such as **musk xylene** and **musk ambrette**, are particularly bad in this respect. Musk ambrette has now been banned as a perfumery material owing to its ability to cause severe skin sensitization in the presence of sunlight. Similar sensitization occurs with **bergamot oil** and with **cumin oil**.

Products known as **UV absorbers**, similar to the products used in sunscreen preparations, can be added to alcoholic lotions that almost eliminate this type of discoloration. Some companies use them as a matter of course in the manufacture of their finished perfumes and toilet waters. But they should not be added directly to the perfume compound. Many of the problems of discoloration in soap, for example, as by **vanillin**, are accelerated by exposure to light.

The Effect of Temperature

Heat is a form of energy that affects the rate at which chemical reactions take place. The shelf life of a product can therefore be considerably reduced by exposure to elevated temperatures. Odor deterioration and discoloration will develop much faster under these conditions, and this is exploited in the accelerated testing of perfumes for stability in final products. Some products are manufactured at quite high temperatures, and perfumers have to keep this in mind in the formulation of their fragrances.

Acids and Alkalies—pH Values

Many product bases that the perfumer has to deal with may be either acidic or alkaline, and as we have already seen, under such conditions a number of materials such as esters, aldehydes, and ketones are likely to be unstable. The severity of the problem depends on the strength and type of acid or alkali concerned. For instance, hydrochloric acid and citric acid are examples of a strong and a weak acid, respectively, and sodium hydroxide and ammonia of a strong and a weak alkali. One way commonly used to express the degree of acidity or alkalinity is by means of the **pH value**. This is a numerical scale from 1 to 14 in which pH 1 is the most acidic and pH 14 the most alkaline. At pH 7 the product is said to be neutral.

HOSTILE BASES AND STABILITY TESTING

As has already been mentioned, many of the functional products with which perfumers are concerned are weakly or strongly hostile toward a wide range of perfumery materials. Household products such as bleaches are powerfully alkaline, whereas those that remove limescale are strongly acidic. Modern detergent powders contain active ingredients that present perfumers with yet another set of problems. Liquid detergents may selectively lower the volatility and thus the performance of different materials, and minerals such as talcum powder may contain traces of heavy metals. Indeed it is difficult to think of any type of base that does not in some way come between perfumers and their creations. Perfumers must always be working within the limitations set by the bases with which they are concerned.

Although it is possible to some extent to predict the stability and performance of certain types of perfumery material in any base—for example, the instability of esters in acid media—perfumers must usually rely on the experience gained from the testing of individual perfumery materials. This is a laborious undertaking and one that occupies much of the time of both the perfumer and the application chemist. As has been mentioned earlier, such testing is normally carried out at elevated temperature as well as at ambient temperature and in the cold.

Assessing the results of such tests requires considerable experience. It is important, for example, to differentiate between the stability of a product and its performance. A material smelled in a sample of dry detergent powder may show poor performance and yet be stable, making a valuable contribution to the performance of a perfume on washed fabric. Its failure to perform in the base, even on cold storage, should not necessarily be attributed to instability.

Similarly a weakly smelling material may, at elevated temperatures, be masked by the smell that has develolped in the base itself. In many cases where a customer has complained about the deterioration of the product this can be put down to the deterioration of the base itself rather than to that of the perfume. It is one of the golden rules of stability testing always to begin by testing the unperfumed base.

Packaging

The importance of packaging has already been mentioned in connection with the problem of oxidation. In most cases where a problem arises due to the faulty design of the packaging or the injudicious selection of component materials there is little that the perfumer can do other than to identify the true cause of the problem.

Various grades of plastic, for example, differ widely in their permeability to perfume. Low-density polyethylene is far more permeable than the high-density variety; PVC, an excellent material in many ways, owing to its toughness and comparative impermeability to perfume compounds, can often be the cause of odor contamination due to the presence of plasticizers used in its manufacture. For environmental reasons PVC, which contains chlorine in its molecular structure, is rarely used today.

Aerosols can often develop a powerful sulphurous rubber smell due to the quality or age of the rubber gasket. Soaps can pick up the smell of their wrapping material such as cardboard or even the smell of a plastic container. So often these problems are blamed on the perfume, and it becomes the job of the perfumer or application chemist to identify the fault correctly.

Discoloration in Soap

Owing to their natural color many perfumery materials such as oakmoss and Schiff bases cause primary discoloration in soap. Their use in white soap base must therefore be restricted. Other materials may cause discoloration with aging either due to reaction with the soap base (oxidation, etc.) or in the presence of light. The use of some of these materials may be justified even in white soaps at low concentrations, and at much higher levels in colored soaps, depending on the actual color and intensity of the final product.

Pink soaps are often as difficult as white, while yellow soaps will allow the use of yellowing materials such as citral and Schiff bases. Similarly beige and brown colored soaps will permit the use of certain amounts of nitromusks and eugenol. The quality of the soap base is important both in its effect on general stability, for example, with aldehydes, and on certain types of discoloration. The presence of iron, for example, can cause problems with salicylates, although most modern soaps contain sequestering agents that effectively inactivate heavy metals.

Unfortunately, some customers insist on white soap stability even when the final product is to be colored. The following is a list of common materials that may cause discoloration:

Materials Causing Discoloration in Soap

NITROGEN-CONTAINING MATERIALS

Nitromusks	Photodiscoloration
Musk ambrette	
Musk ketone	

NITROGEN-CONTAINING MATERIALS

Musk xylene	(Excellent in soap when color permits)
(Musk tibetine does not discolor)	
Indol	Severe discoloration even in traces
Indolene	
Skatol	
(Indolal and lacktoskatone do not discolor)	
All anthranilates	Severe yellowing due to Schiff base formation with free aldehydes in the base
(Quinolines can cause a problem, but iso-butyl quinoline is stable at normal concentrations)	

AROMATIC ALDEHYDES

Vanillin	Severe discoloration
Ethyl vanillin	Use at low concentrations up to 0.1%
Heliotropin	Use in moderation
(Anisaldehyde does not discolor)	
Cinnamic aldehyde	Discolors
Amyl cinnamic aldehyde	May cause discoloration at high levels
(Hexyl cinnamic aldehyde does not discolor)	
Helional	Photodiscoloration

OTHER ALDEHYDES

Some aldehydes can cause discoloration due to polymerization in alkaline soap base.

Citral	Yellow–brown discoloration

PHENOLS

Eugenol	Brown discoloration
Isoeugenol	Severe darkening to black
Thymol	Darkening

NITROGEN-CONTAINING MATERIALS

Naturals containing phenols cause discoloration (e.g., oakmoss, birch tar, castoreum, and oils containing eugenol)

Most of the esters of eugenol and isoeugenol, as well as benzyl isoeugenol, cause discoloration.
(Methyl eugenol and methyl isoeugenol are stable)

Salicylates	Discolor in the presence of iron

OTHER MATERIALS known to cause discoloration

Estragol
Dimethyl hydroquinone
Nerolin yara yara
Bromstyrol

19

The Physical Basis of Perfumery

The properties of perfume materials are intimately related to their chemical constitution, but the mechanisms whereby chemical structure leads to odor perception involve, in crucial ways, a physical phenomenon: the mutual attraction forces between molecules. These forces determine the rate of evaporation of odor materials from solutions or surfaces, they are the basis of fixation and substantivity, they explain why the odor quality of mixtures varies depending upon the solvent or base in which they are incorporated. They are also involved in the very process of odor perception, in the contact between the odorant molecule and the receptor cell. Moreover they are at the heart of distillation, extraction, solubility, and the mechanism of chromatography. In this chapter, we will briefly discuss the physical basis of some of these phenomena, showing also how the attraction forces between molecules are related to their chemical structure.

EVAPORATION AND VOLATILITY

Perfume materials, in the form in which we apply them to the smelling blotter, mix them, weigh them, and incorporate them into products are usually liquids, occasionally also pastes or solids. The products in which they are incorporated are usually liquids, emulsions, gels, or solids.

Odor perception takes place only when molecules of the odorous material enter the nasal cavity and physically touch the olfactory receptors located there. In nearly all instances (the exception being finely dispersed aerosols), this means that perfume materials have to pass into the **vapor phase** in order to be perceived. The process of **evaporation** therefore is of central importance to perfumery. The **volatility** of perfume materials, that is, their readiness to pass into the vapor state, is one of their most important properties from the perfumer's point of view (Appell 1964). In fact two famous perfumers have proposed methods of perfume composition based upon volatility (Poucher 1955; Carles 1961).

CHEMICAL STRUCTURE AND VOLATILITY

As a general rule the larger the molecular size—that is, the higher the number of carbon atoms—the lower is the volatility of the compound. For example, the **monoterpenes** (C10) are more volatile, and therefore less persistent, than the **sesquiterpenes** (C15).

However, the presence in a molecule of a functional group containing oxygen has the result of greatly reducing the volatility. For example, the three hydrocarbons, ethane (C2, a gas), benzene (C6), and cedrene (C15), are all very much more volatile than their corresponding alcohols, ethyl alcohol, benzyl alcohol, and cedrenol. Here again, as was seen in relation to odor, the smaller the molecule, the greater is the effect of the functional group upon volatility. Two functional groups within the same molecule tend to lower the volatility even further, as has already been pointed out in the case of hydroxycitronellal.

This effect of functional groups containing oxygen in reducing the volatility is due to the polarization of the electrical charge (the electron configuration) within the molecule. Such **polar molecules** are mutually attracted, and said to be **associated**, thus reducing their readiness to separate and lowering the volatility. The degree of association due to polarization varies from one type of functional group to another. Aldehydes, ketones, and esters, for example, which contain a carbonyl group, C=O, are less strongly associated than either alcohols or acids in which the presence of a hydroxy, —OH, group results in a further type of association known as **hydrogen bonding**. In alcohols this produces the formation of a weakly linked chain structure. In acids, where a hydroxyl group and a carbonyl group together make up the characteristic carboxylic group, $—\overset{\overset{\displaystyle O}{\|}}{C}—OH$, an even stronger linkage is formed between pairs of molecules, or **dimers**. Although the nature of these hydrogen bonds is too complicated to be discussed here in detail, they may be represented simply as follows:

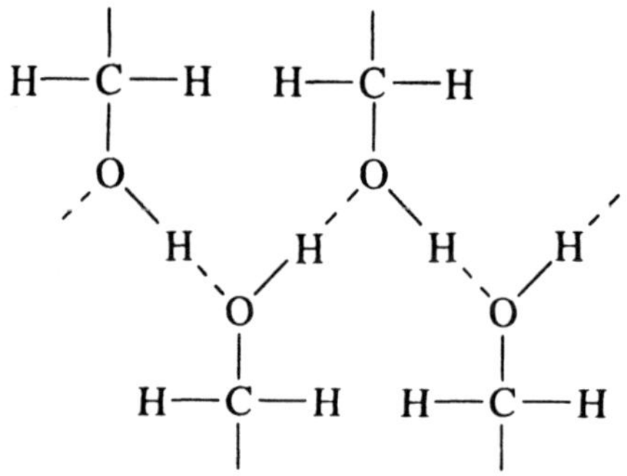

Hydrogen bonding in alcohols

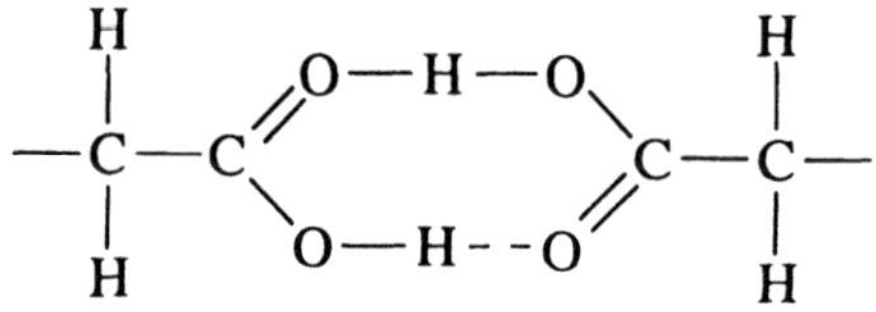

A carboxilic dimer

These links are broken when the temperature of the material reaches boiling point, or when the material evaporates. So what is smelled are the individual molecules.

Association, due to polarization and hydrogen bonding, is of special significance to perfumers when it comes to the **fixation** of perfumes, and in the formulation of perfumes for such products as fabric detergents and conditioners where **substantivity** is a major requirement. The polar characteristics of perfumery materials are also of importance for their separation by **gas chromatography** when using certain types of column.

The volatility of esters is also worth considering. As was mentioned previously, the esters are formed by the combination of alcohols and acids. The resulting molecule has one weakly polarizing carbonyl group. The esters therefore tend to be more volatile than either alcohols or acids of comparable molecular size. However, the combined molecular size of an ester may be far greater than one or both of its parent molecules, and this will affect its relative volatility.

For example, ethyl phenylacetate is far more volatile than phenylacetic acid, although the molecular size is greater. Similarly phenylethyl acetate is slightly more volatile than phenylethyl alcohol. However, phenylethyl phenylacetate is very much less volatile than phenylethyl alcohol and rather less so than phenylacetic acid. This can be related to the following structural formulas:

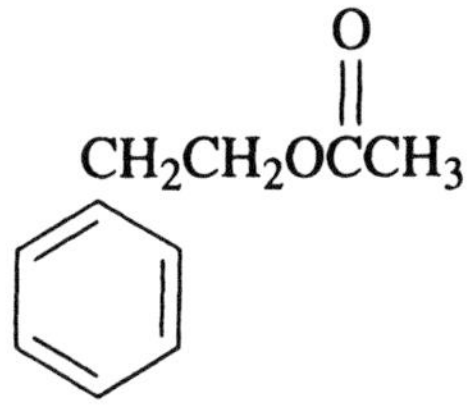

$C_6H_5CH_2CH_2OC(=O)CH_3$

Phenylethyl acetate

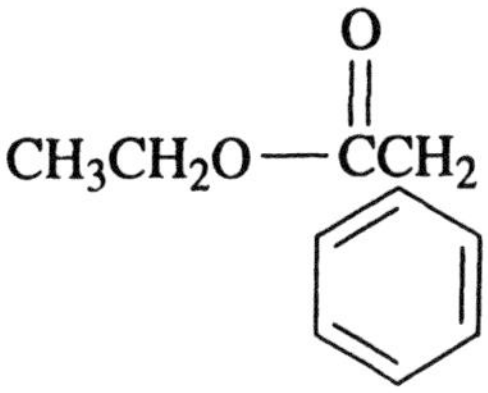

$CH_3CH_2O—C(=O)CH_2C_6H_5$

Ethyl phenylacetate

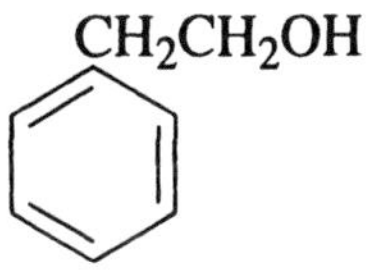

$C_6H_5CH_2CH_2OH$

Phenylethyl alcohol

$C_6H_5CH_2CH_2O—C(=O)—CH_2C_6H_5$

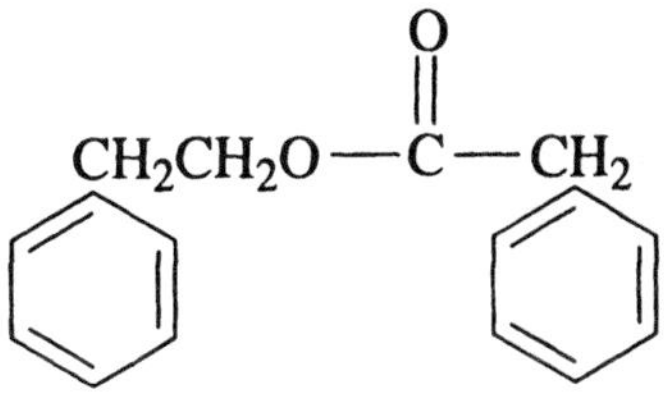

Phenylethyl phenylacetate

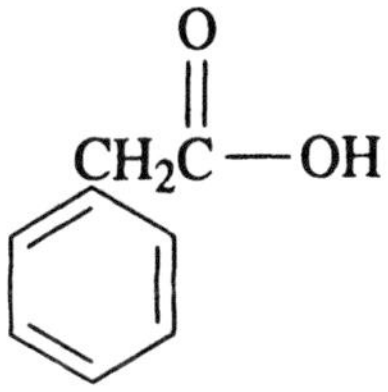

$C_6H_5CH_2C(=O)—OH$

Phenylacetic acid

Although esters, like acids, have a carbonyl group attached to another oxygen atom, the second oxygen atom in this case is linked on the other side to another carbon. This type of linkage, characteristic of the **ethers**, is nonpolar, and has little effect therefore on volatility.

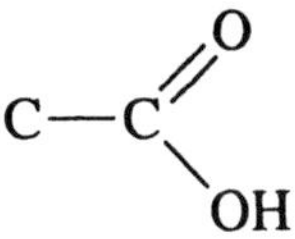

C—C(=O)—OH

Acid

C—O—C

Ether

C—O—C(=O)—

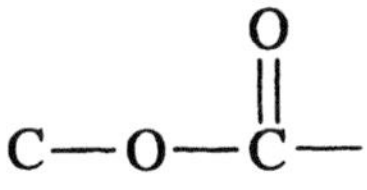

Ester

This is further demonstrated by the two ethers phenylethyl methyl ether (pandanol) and phenylethyl iso-amyl ether (anther), both of which are far more volatile than phenylethyl alcohol, although having a greater number of carbon atoms:

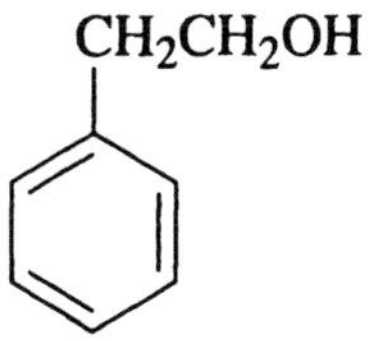

$C_6H_5CH_2CH_2OH$

Phenylethyl alcohol

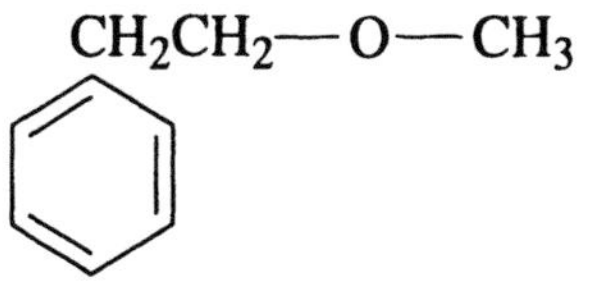

$C_6H_5CH_2CH_2—O—CH_3$

Pandanol

$C_6H_5CH_2CH_2OCH_2CH_2CH(CH_3)_2$

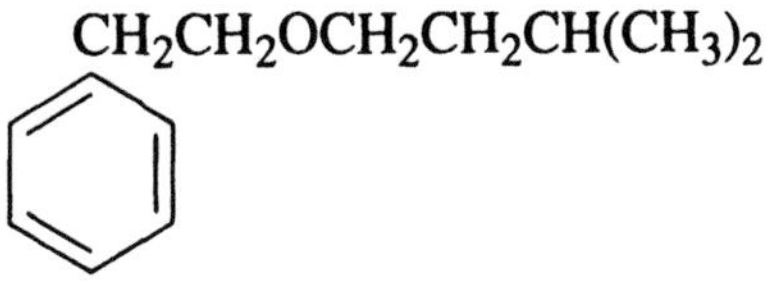

Anther

Rose oxide, although a monoterpene derivative with 10 carbon atoms, is also very volatile, being a cyclic ether with a six-membered pyran ring but with no polar functional group. Very large molecules such as Galaxolide, with 18 carbon atoms in a complex three-ring structure, can be very persistent despite having no functional groups. This is entirely due to the size of the molecule.

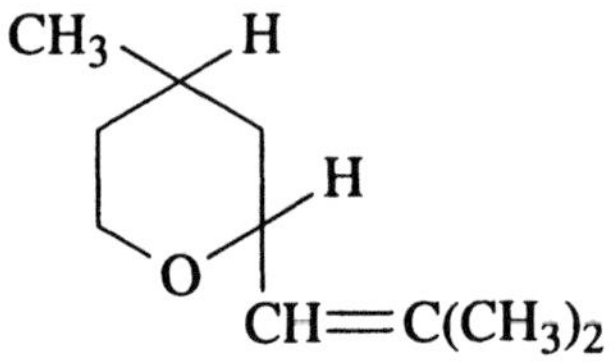

CH_3 H H O $CH{=}C(CH_3)_2$

Rose oxide

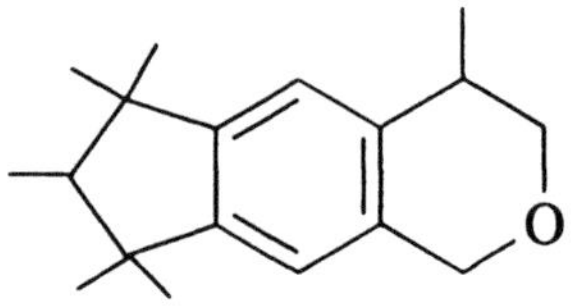

O

Galaxolide

In summary, large molecules tend to be more persistent than smaller ones, though this may be modified by the presence of functional groups, which cause polarization and hydrogen bonding. For an equivalent molecular size, hydrocarbons and ethers are more volatile than aldehydes, ketones, and esters, which are in turn more volatile than alcohols and acids.

Most of the examples given so far have been chosen, for the sake of simplicity, from compounds whose functional groups are made up of carbon, hydrogen, and oxygen. Those containing nitrogen, the **nitrogenous** compounds, although less numerous in perfumery, are still of great importance. These will be seen from the lists of structural and functional groups. (See Appendixes A and B.) A useful measure of the volatility of a substance is its vapor pressure at room temperature. Tables 13.1 through 13.3 provide this for a number of aroma chemicals.

SOLUBILITY AND ODOR PERFORMANCE

Forces of mutual attraction and hydrogen bonding occur not only among the molecules of any given material but also between the different kinds of molecules that make up mixtures and solutions. They directly affect the solubility of perfume materials in different solvent systems, their odor performance in these systems, and the odor character of the solution.

The solubility of substances in different solvents is determined by the **relation** between the attraction forces among the molecules of that substance and the attraction forces that occur between the molecules of the solvent. If the two are similar in strength and in kind, solubility is high. If they are very different, solubility is poor. The alchemists said, "like dissolves like"; today we say that polar solvents are good solvents for polar substances and poor ones for nonpolar substances, and the reverse holds for nonpolar solvents. The main functional groups and solvents of interest to perfumers, in order of decreasing polarity, are as follows:

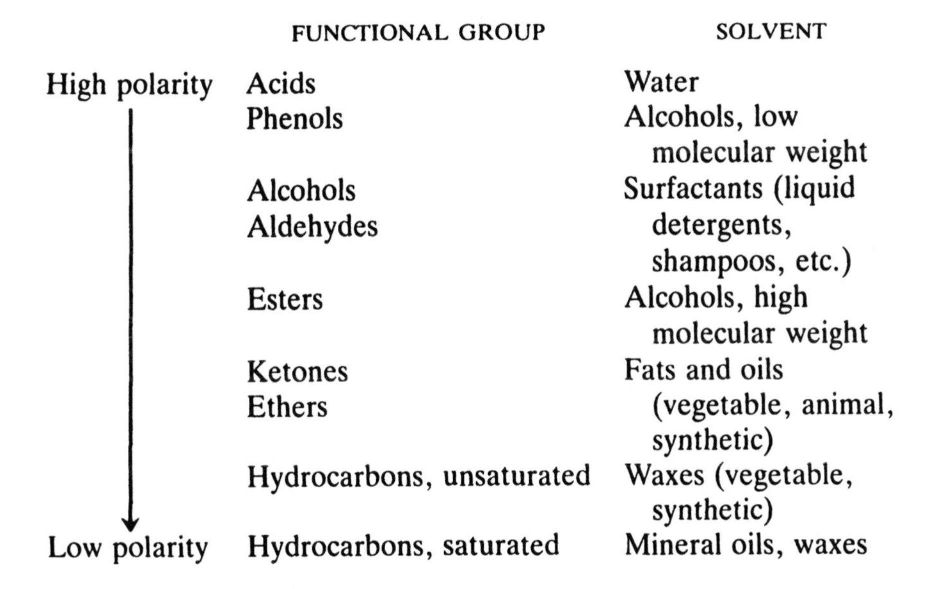

	FUNCTIONAL GROUP	SOLVENT
High polarity	Acids	Water
↓	Phenols	Alcohols, low molecular weight
	Alcohols	Surfactants (liquid detergents, shampoos, etc.)
	Aldehydes	
	Esters	Alcohols, high molecular weight
	Ketones	Fats and oils (vegetable, animal, synthetic)
	Ethers	
	Hydrocarbons, unsaturated	Waxes (vegetable, synthetic)
Low polarity	Hydrocarbons, saturated	Mineral oils, waxes

Within any functional group, polarity decreases rapidly as the length of the nonpolar carbon chain increases. The polarity of mixed solvents is intermediate between the respective polarities of individual solvents of which they are composed.

In perfumery practice solubility problems occur only at the extremes of the solvent range. At the high polarity end of the scale, the systems contain high proportions of water such as low-degree alcoholic skin lotions or after shaves, or foam baths and dishwashing liquids (surfactant-water blends) with very low surfactant levels. In the latter, salt is often added to increase viscosity. This addition further increases the polarity of the water and aggravates perfume solubility problems.

Solubility problems in aqueous-alcoholic systems become more pronounced at low temperatures, such as during outdoors storage in cold climates or during air transport. Solubility problems in surfactant-water blends, on the other hand, increase at higher temperatures.

At the low polarity end of the scale, problems may occur in skin oils (e.g., massage oils) containing high proportions of mineral oils or in paraffin candles. Solubility problems never occur in multiple-phase products such as emulsions or soaps, since here the perfume materials have the choice of entering whichever phase they are most soluble in. Strong mutual attraction between solvent and perfume material molecules results in a decreased tendency of the perfume material to evaporate (it is "held back" by the solvent), and hence lowers the odor intensity of that material above the solution.

Since perfume compositions normally are blends of perfume materials that differ widely in their polarity (even single essential oils represent such blends), the different components of a perfume are held back to differing degrees when the perfume is dissolved in a solvent. The patterns of differential holding vary greatly between different solvent systems. As a result, if a given perfume composition is dissolved in two different systems, the odor above the two will be distinctly different (Jellinek 1959).

It can therefore be stated generally that the odor of any perfumed product is affected by the product base in two ways: (1) by the odor of the base itself and (2) by the way the base, through physical attraction forces, affects the odor of the perfume. Sometimes a third factor, chemical decomposition of the perfume by components of the base, enters the picture. This will be considered later.

ODOR PERSISTENCE AND FIXATION

The persistence of a perfume material depends on its volatility: the greater the volatility, the less is the persistence. However, in any given application, persistence also depends on the concentration of the material employed (the higher the dosage, the greater is the persistence) and on the strength of the attraction forces between the **perfume material** and the **base** in which this material has been incorporated, as well as the **substrate** to which it is applied.

Let us, for example, examine the case of a solution of a perfume material applied to a smelling blotter. If we compare the persistence of a 10% alcoholic solution with that of a 1% alcoholic solution, both dipped to the same height, we will find the more concentrated solution to be more persistent. If we compare two 1% solutions of the same material, one in a volatile solvent such as ethyl alcohol and one in a nonvolatile solvent such as diethyl phthalate or dipropylenene glycol, on blotters, again taking care to apply the same amounts of both solutions, we may find the solution in the nonvolatile solvent to be more persistent due to the attraction forces between the solvent and the perfume material. Ethyl alcohol evaporates so quickly from the blotter that such forces soon cease to be effective in the case of the alcoholic solution.

In the case of weak odorants we may observe a reverse effect. Again the perfume material is held more strongly by the nonvolatile solvent, but this also diminishes its intensity in the air over the blotter. Thus it may happen that the material, although still physically present, is no longer readily perceptible because it has become too weak.

Coming back once more to the alcoholic solutions: After the alcohol has evaporated, there are still attraction forces of the dipole-dipole or hydrogen bonding kind at work that slow down the evaporation of the perfume material. These originate in part from interactions between the molecules of the different materials contained in the perfume. Perfumers may use materials of low volatility in their perfumes intending thereby to slow down the evaporation of more volatile perfume components. This is the practice commonly referred to as **fixation** (Sturm and Mansfeld 1976; Jellinek 1978).

ADHESION AND SUBSTANTIVITY

Attraction also occurs between the paper of the smelling blotter and the perfume materials. The attraction forces between the perfume and the solid support upon which it has been deposited are called **forces of adhesion**, and the degree of adhesion is often designated by the term **substantivity**. Adhesion plays a major role in the persistence of perfume on the skin, of the perfume from shampoos or conditioners on the hair, of detergent and fabric softener perfumes on textile fibers.

In the latter application the dependence of adhesion upon the nature of the support can readily be observed. Under comparable conditions of application, perfumes are far more persistent upon wool fibers than upon nylon; cotton occupies an intermediary position. This is because the molecular structure of nylon, as distinct from wool or cotton, offers little or no opportunity for hydrogen bond formation. Conversely, there are great differences in the degree of adhesion of different perfume materials to a given support. This is the principle upon which both liquid and vapor chromatography are based.

The persistence of perfume materials on human skin has been the subject of some study (Wells 1960; Jellinek 1964). But, considering the importance of evaporation, and of the attraction forces between molecules, to the performance of perfumes in all conceivable applications, it is surprising how little attention has been paid to these subjects by the perfumery profession at large.

DIFFUSION

The diffusion of gases is a physical phenomenon. It would be reasonable to think that the rate of diffusion of the vapors of perfume materials in air should have a major effect on their odor diffusion and volume and hence be a matter of great practical concern to the perfumer. In practice this is not the case, for two reasons.

The rate of diffusion of a substance is proportional to the square root of its molecular weight. Since the common perfumery chemicals span a fairly narrow spectrum of molecular weights, this means that the lightest common materials travel about 1.5 times as fast as the heaviest ones. This is not a major difference.

Even more important is the fact that under normal conditions of perfume use, there is so much turbulence in the air that the odorant molecules are distributed through turbulent flow patterns rather than straight-line diffusion. In such flow patterns, the specific nature of the molecule plays a small role indeed.

20

Psychophysics and Perfumery

Psychophysics explores the relationships between measurable characteristics of stimuli and their perception. In the case of odor as a stimulus, the only two quantitatively measurable aspects, at our current state of knowledge, are concentrations and the composition of mixtures. Hence the psychophysics of odor deals with measuring the perception of given odorants at different concentrations and with the perception of mixtures.

In the brief discussion that follows, we will again restrict ourselves to those aspects of the subject that have direct and practical application to perfumers' work. They relate only to relations between concentration and perception, for although considerable research has been and is being conducted upon the perception of mixtures of odorants, this research has not, so far, generated findings of immediate interest to perfumers.

THRESHOLDS

A fundamental concern of psychophysicists working on olfaction has been the determination of the **odor thresholds** of a broad array of substances. A distinction must be made between the **detection threshold**, which is the lowest concentration at which significant detection takes place that some odorous material is present, and the **recognition threshold**, which is the lowest concentration at which an odorant can be recognized for what it is.

Because thresholds differ from person to person and can vary from one occasion to another even within the same person, reliable threshold values can be obtained only as averages taken from sizable groups of measurements. Moreover reports of threshold concentrations are meaningful only if they specify under what conditions they were established, since the threshold concentration of any given substance depends a great deal upon the precise condition under which it has been established.

For instance, as is shown in Table 20.1, very different threshold values are obtained if the concentration of the substance has been measured in a solution the head space of which is smelled, or directly in thc inspired air. In the case of solutions the specific nature of the solvent makes a great deal of difference (compare Table 20.2). In two-solvent systems such as emulsions, the higher of the two thresholds prevails, since the odorant distributes itself between the two phases in such a way that it resides predominantly in the phase in which it is most soluble, which is usually the one in which its threshold is highest. Thresholds based on the concentration in air are the most meaningful values, since they are independent of the medium in which the odorant was dissolved, but they are also the most laborious to measure, requiring an elaborate apparatus named **olfactometer**.

In the normal applications of perfumes, concentrations come into play that lie far above the threshold values. Nevertheless, when used in conjunction with vapor pressure data, the threshold values of individual odorants can give the perfumer helpful information about their performance. Substances with low threshold values are generally more potent in odor than substances with comparable vapor pressure (volatility) and higher thresholds.

Table 20.1 shows the detection thresholds of a number of perfume materials in air and in water. Note the tremendous range (from 0.002 parts per billion for beta-damascenone to 10,000 parts per billion for phenylacetic acid—both taken in water solutions), the large difference between optical isomers of the same substance (e.g., Nootkatone and alpha-damascone), and the large differences in thresholds reported by different investigators (e.g., benzaldehyde and vanillin). In substances with relatively high water solubility such as vanillin and ethyl vanillin, benzaldehyde, phenylethyl alcohol, and phenylacetic acid, the thresholds in water are very much higher than in air. In poorly water-soluble substances such as pinene and the macrocyclic musk cyclopentadecanolid, the reverse is true. The relative thresholds of a substance in different solvents indicate its performance in different application environments. Substances whose thresholds in water solution are much

TABLE 20.1 Detection Thresholds of Some Odorants (in ppb)

Substance	Threshold in Air[a]	Threshold in Water[b]
Alcohol C8	5.8	110–130
Alcohol C9	71	50
Alcohol C10	18	
Alcohol C11	68	
Alcohol C12 (lauric)	13	
Aldehyde C8	1.3	0.7
Aldehyde C9	2.2	1.0
Aldehyde C10	0.9	0.1–2.0
Aldehyde C11 (saturated)	1.7	
Aldehyde C11 (undecylenic)	2.0	5.0
Aldehyde C12 (lauric)		2.0
(+)-Ambrox		2.6
(−)-Ambrox ("Ambroxan")		0.3
DL-Ambrox ("Synambran")		0.6
Amyl salicylate	3.5	
Androstenone	1.1	
Anethole	7.1	
Anisaldehyde	33	
Benzaldehyde	42	350–3500
Benzyl acetate	145	
Borneol	2.1	
Bornyl acetate	14	75
Camphor	51	
(−)-Carvone	22	50
Cineole (Eucalyptol)	16	12
Cinnamaldehyde	2.6	
Cinnamic alcohol	1.2	
Citral a (Geranial)	3.2	32
Citral	7.4	30
Citronellal	11	
Citronellol	7.1	
(−)-Citronellol		4
Coumarin	0.7	
Cyclocitral, beta		5
Cyclopentadecanolid	102	1–4
Damascenone, beta		0.002
(+)-α-Damascone		100
(−)-α-Damascone		1.5
Decalactone, gamma		11
Decalactone, delta		100
Dodecalactone, gamma		7
Diphenylmethane	23	
Ethyl alcohol (Ethanol)	29,000	
Ethyl benzoate	28	60
Ethyl phenylacetate		65
Ethyl vanillin	0.07	100
Eugenol	11	6–30
Farnesol		20
Fenchone	93	
Geraniol		40–75

TABLE 20.1 (*Continued*)

Substance	Threshold in Air[a]	Threshold in Water[b]
Geranyl acetate		9
Geranyl isobutyrate		13
Geranyl propionate		10
Guaiacol	1.0	
Heliotropine	4.8	
Heptalactone, gamma		400
Hexalactone, gamma		1600
cis-3-Hexenol		70
Indole	0.03	140
Ionone, alpha	0.06	
Ionone, beta	2.6	0.007
Limonene	440	
Linalool	54	6
Linalyl acetate	8.9	
Maltol (Corps praline)		35,000
Menthol	42	
Menthone		170
p-Methylacetophenone		0.027
Methyl anthranilate	1.1	
Methyl benzoate	11	
Methylcyclopentenolone		300
Methyl eugenol		820
Methyl heptine carbonate	0.8	
Methyl nonyl ketone	22	7
Methyl salicylate	44	40
Musk ambrette	0.002	
Musk xylene	0.06	
Nerol	0.2	300
Nonadienal, *trans*-2, *cis*-6	0.01	0.01
Nonadienal, *trans*-2, *trans*-4	0.04	0.09
2-Nonenal		0.08–0.1
(+)-Nootkatone		0.8–1
(−)-Nootkatone		600
Phenylacetaldehyde		4
Phenylacetaldehyde dimethyl acetal	21	
Phenylacetic acid	1.3	10,000
Phenylethyl alcohol	17	750–1100
Pinene, alpha	690	6
Pinene, beta		140
Raspberry Ketone		100
Skatole	0.6	
Styrallyl acetate	7.2	
Terpineol, alpha	37	300–350
Thymol	15	
Vanillin	0.03	20–200

[a]From Devos et al. (1990). This is a compilation from 372 references. For each material a "standard value" was derived by a complex mathematical treatment of the available values from these sources.
[b]From Leffingwell and Leffingwell (1991). This also is a compilation from multiple sources. Where values for one substance from different sources showed marked differences, the highest and lowest value reported are listed.

TABLE 20.2 Ratio of Flavor Threshold* of Aliphatic Aldehydes in Different Solvents

	Approximate Mole Ratio	
Substance	Threshold in Peanut Oil / Threshold in Water	Threshold in Paraffin Oil / Threshold in Water
Aldehyde C_1 (propionaldehyde)	1	6
Aldehyde C_4 (butyraldehyde)	3	10
Aldehyde C_5 (hexyl aldehyde)	10	20
Aldehyde C_8 (octyl aldehyde)	200	100
Aldehyde C_{10} (decyl aldehyde)	1000	400
Aldehyde C_{12} (lauric aldehyde)	10,000	2000
Aldehyde C_{14} (myristic aldehyde)	135,000	9000
Aldehyde C_4 unsatd.	1	0.5
Aldehyde C_9 unsatd.	13	70
Methyl hexyl ketone	8	

Source: Lea and Swoboda (1958).

lower than in oily solutions are more odor effective in applications where no oil phase is present (e.g., in gels or water-surfactant systems such as shampoos) than in products containing an oil phase, such as emulsions or soap. As shown in Table 20.2, the higher fatty aldehydes are examples of this type. On the other hand, substances whose thresholds in oil and in water are not so far apart (e.g., the lower aldehydes, especially if unsaturated) vary little in their performance whether or not oil is present.

CONCENTRATION AND INTENSITY

Normally, the perfumer is interested in the performance of odorants at levels well above their threshold concentrations. How strong are they or, to put it more scientifically, how intensive is their odor perceived to be? The answer obviously depends on their concentration in the perfume and in the finished product: the greater the concentration, the higher is the intensity.

The relationship between intensity and concentration is, however, not the same for all odorants. With some odorants the intensity increases distinctly as the concentration is increased, with others the change in intensity is far less marked. The dependence of intensity

*The flavor thresholds in Table 20.2 are perceived by the nose when the solutions are placed in the mouth. It is reasonable to assume that the *threshold ratios* would be similar upon smelling the solutions directly.

TABLE 20.3 Slopes of Selected Substances

Substance	Slope
Amyl acetate	0.21
Anethole	0.24
Benzyl acetate	0.33
Citral	0.25
Coumarin	0.25
Decyl alcohol	0.12
Eugenol	0.40
Geraniol	0.26
d-Menthol	0.36
Methyl anthranilate	0.49
Musk xylene	0.30
Phenylacetic acid	0.18
Phenylethyl alcohol	0.28
Terpineol alpha	0.39

Source: Patte et al. (1975).

upon concentration can be indicated by a parameter which, for reasons that will not concern us here, is called the **slope of the psychophysical function** or simply the **slope.*** The values for the slopes of a few perfume materials are shown in Table 20.3.

A relatively high value for the slope indicates a strong dependence of intensity upon concentration. With a substance having a slope of 1, a tenfold increase (or decrease) in concentration would result in a tenfold increase (or decrease) in intensity. With a substance having a slope of 0.40, a tenfold change in concentration results in an intensity change by a factor of only $10^{0.40} = 2.5$. With a substance with slope 0.20, the intensity would change by a factor of only $10^{0.20} = 1.6$.

In applications where fragrance tenacity and diffusion are desired, odorants with a low slope are more effective than those with higher slopes, other things (e.g., volatility and threshold) being equal (Jellinek 1979). This is because the concentration of odorants in the air near a perfumed product, object, or person always decreases as time elapses after application of the perfume (or exposure of the product), and it is always less at a greater distance than near the product, object, or person. Now, if this lesser concentration in the air results in a but slight weakening of intensity (as it does with substances of low slope), there is good tenacity and good diffusion.

*The notion of the slope as a constant, characteristic of a substance assumes the validity of Stevens's law (Stevens 1957) for odors. This is still a matter of some controversy. For example, Sauvageot (1987) holds the slope to be "characteristic of the stimulus-subject couple." Certainly the value for the slope depends upon the method by which it is determined.

Low-slope odorants typically do not seem very powerful when examined undiluted on a blotter and for this reason their value may be overlooked in the initial screening of new perfumery materials. They show their mettle only when used in actual applications. We suspect that materials such as the synthetic musks, Hedione, and the damascones are, among others, of this kind.

Unfortunately, the determination of the slope is a laborious undertaking beset with pitfalls (Hyman 1977). It may well be that values given in Table 20.3 are in error. However, a rough measure of the dependence of intensity upon concentration may be obtained by examining dilution series of odorants upon the blotter. Those materials that are still clearly recognizable at 0.1%, even though they did not appear to be particularly powerful at 10% or 100%, are the ones of special interest.

SLOPES AND ODOR VALUES

The odor value (OV) concept and its uses were discussed in Chapter 13. The odor value, being defined as

$$OV = \frac{\text{Actual head-space concentration of odorant}}{\text{Threshold concentration of odorant in air}}$$

would be a true indication of odor strength if the odorant's intensity varied in direct proportion to its concentration in the air, i.e., if the slope of all substances were equal to 1. As Table 3 shows, this is clearly not the case. To obtain a true value of intensity at levels above threshold, we should take the slope of the substance in question into account. The true odor value of an odorant, OV_t, could be defined by

$$OV_t = OV^{\text{odorant slope}}$$

In the calculations of cost effectiveness of different odorants in a fabric softener application (Chapter 13), we should really have used the OV_t values of the different materials. Not having reliable values for their slopes, we could not calculate the OV_t values and instead used what we called OV' values. These were calculated from the OV values assuming that the slopes of all materials involved is 0.35, certainly not a true assumption but a better one than the assumption of a slope of 1.0 for all materials on which the OV values are based.

The true OV_t values are a useful potential tool for measuring the cost effectiveness of perfumery materials and for creating fragrances of optimal cost effectiveness in use. In quantitative head-space chromatography, we now possess a good technique for establishing the vapor pressures of odorants under relevant conditions. It is to be hoped that before too long the perfumery industry will undertake the effort to establish reliable values for the thresholds and the slopes of a wide range of perfumery materials.

Appendixes

Appendix A

Structural Groups

SUFFIXES

The suffix -yl is normally used in conjunction with alcohols and their derived esters. This is the form given in most of the samples below. The -ic suffix is normally used in conjunction with aldehydes, and for acids. In some cases, however, the two forms are used alternatively, for example, as with cinnamyl or cinnamic alcohol.

The suffix -ate is always used to indicate the ester radical derived from the corresponding acid.

SYNONYMS

In many cases a number of different synonyms occur. Only the most commonly used are given in the following table.

STRUCTURAL GROUPS DERIVED FROM THEIR CORRESPONDING ALCOHOLS		STRUCTURAL GROUPS DERIVED FROM THEIR CORRESPONDING ACIDS	
For example:		For example:	
Methyl	CH_3—	—OOC—H	Formate
Methyl alcohol	CH_3OH	H—COOH	Formic acid

The Aliphatic Series

Ethyl	CH_3CH_2—	—OOC—CH_3	Acetate
Propyl	$CH_3CH_2CH_2$—	—OOC—CH_2CH_3	Propionate
Iso propyl	$(CH_3)_2CH$—		
Allyl	$CH_2{=}CHCH_2$—		
Butyl	$CH_3(CH_2)_2CH$—	—OOC—$(CH_2)_2CH_3$	Butyrate
Iso butyl	$(CH_3)_2CHCH_2$—	—OOC—$CH(CH_3)_2$	Iso butyrate
Amyl	$CH_3(CH_2)_3CH_2$—	—OOC—$(CH_2)_3CH_3$	Valerianate
Iso amyl	$(CH_3)_2CHCH_2CH_2$—	—OOC—$CH_2CH(CH_3)_2$	Iso valerianate
Hexyl	$CH_3(CH_2)_4CH_2$—	—OOC—$(CH_2)_4CH_3$	Caproate
3-Hexenyl	$CH_3CH_2CH{=}CHCH_2CH_2$—	—OOC—$(CH_2)_5CH_3$	Heptoate
Heptyl	$CH_3(CH_2)_5CH_2$—	—OOC—$(CH_2)_6CH_3$	Caprylate
Octyl	$CH_3(CH_2)_6CH_2$—	—OOC—$(CH_2)_7CH_3$	Pelargonate
Nonyl	$CH_3(CH_2)_7CH_2$—	—OOC—$(CH_2)_8CH_3$	Caprinate
Decyl	$CH_3(CH_2)_8CH_2$—		
Undecylic	$CH_3(CH_2)_9CH_2$—		
Undecylenic	$CH_2{=}CH(CH_2)_8CH_2$—		
Lauric, dodecyl	$CH_3(CH_2)_{10}CH_2$—	—OOC—$(CH_2)_{10}CH_3$	Laurate
Tridecyl	$CH_3(CH_2)_{11}CH_2$—		
Myristic	$CH_3(CH_2)_{12}CH_2$—	—OOC—$(CH_2)_{12}CH_3$	Myristate

The Aromatic Series

Phenyl

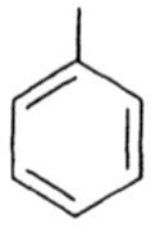

Benzyl — CH_2—

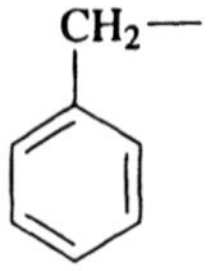

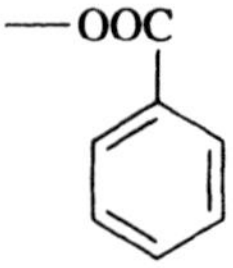

Benzoate — —OOC

Phenylethyl — CH_2CH_2—

Phenylacetate — —OOC—CH_2

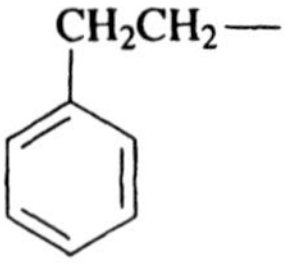

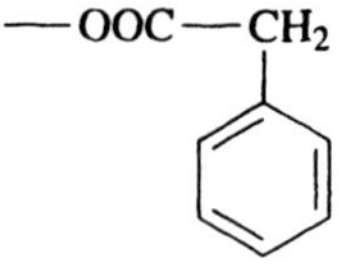

Phenoxyethyl — O—CH_2CH_2—

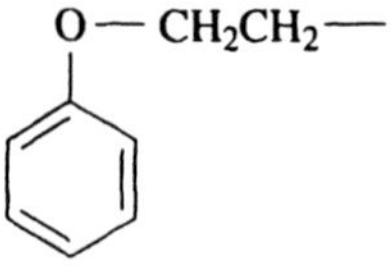

Phenylpropyl hydrocinnamyl — CH_2CH_2CH—

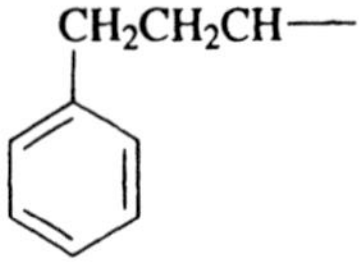

Cinnamyl — CH=$CHCH_2$—

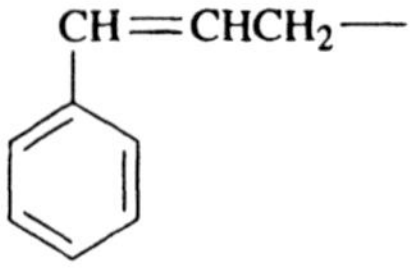

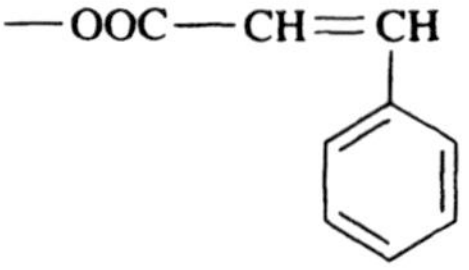

Cinnamate — —OOC—CH=CH

Anisyl — CH_2—, OCH_3

Anisate — —OOC, OCH_3

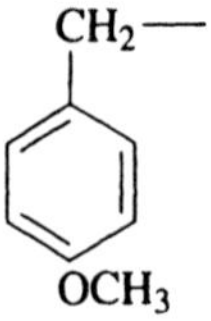

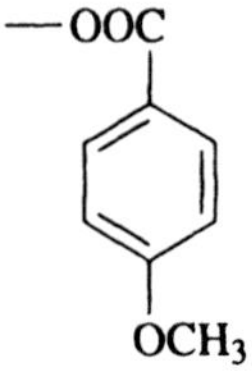

Salicylic — CH_2—, OH

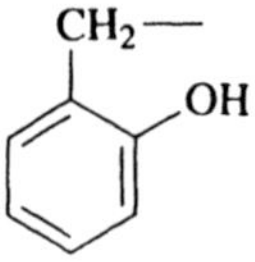

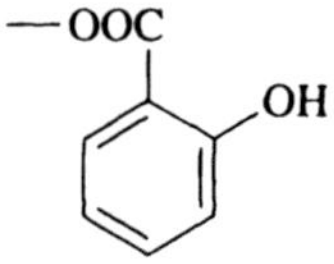

Salicylate — —OOC, OH

para-Cresyl — O—, CH_3

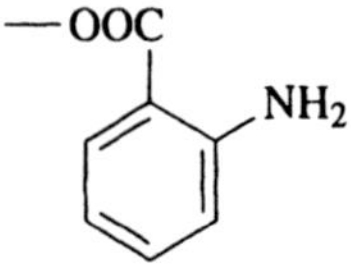

Anthranilate — —OOC, NH_2

The Terpenic Series

(DERIVED FROM THEIR ALCOHOLS)

Geranyl

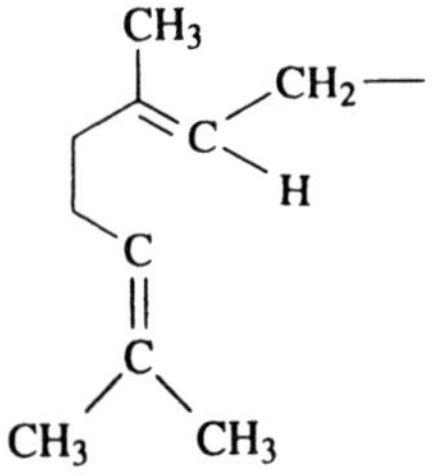

CH_3 $CH_2—$ C H C C CH_3 CH_3

Neryl

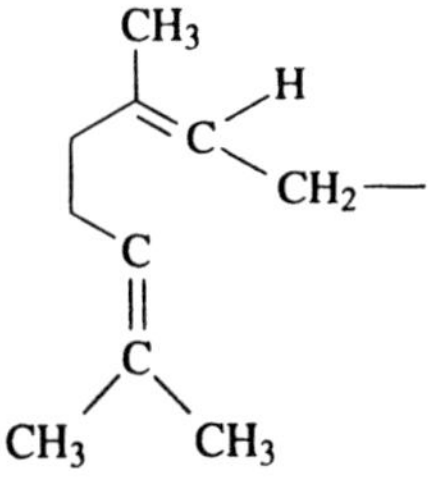

CH_3 H C $CH_2—$ C C CH_3 CH_3

Citronellyl

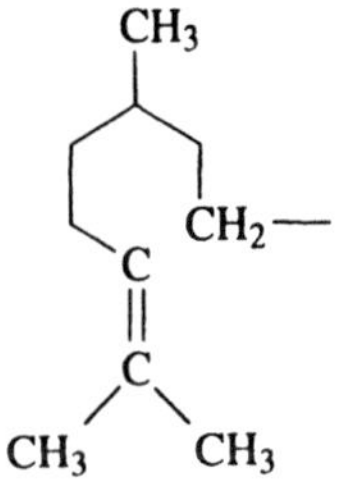

CH_3 $CH_2—$ C C CH_3 CH_3

Linalyl

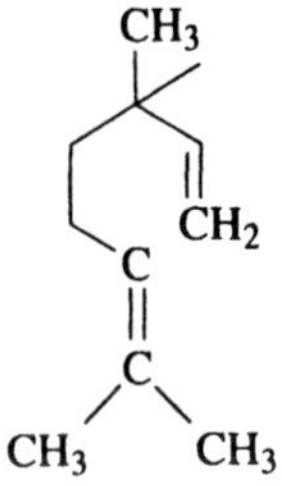

CH_3 CH_2 C C CH_3 CH_3

Terpinyl (alpha)

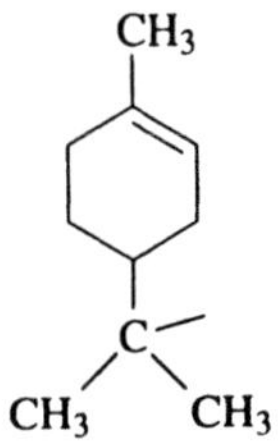

CH_3 C CH_3 CH_3

Iso bornyl

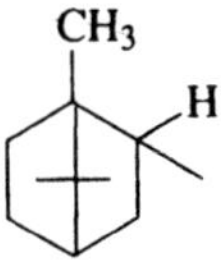

CH_3 H

Cedryl

CH_3 CH_3 CH_3 CH_3

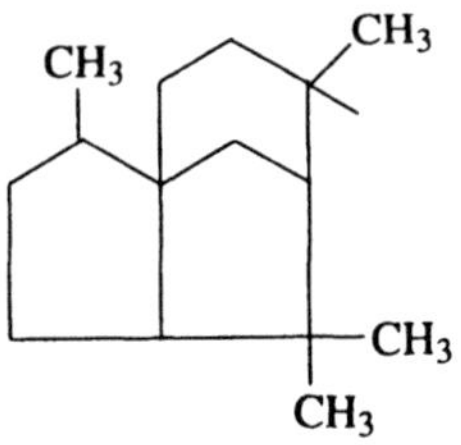

Miscellaneous Groups

Example

Carbinols (secondary alcohols)

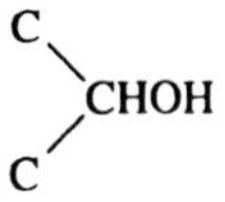

C CHOH C

Methyl benzyl carbinol

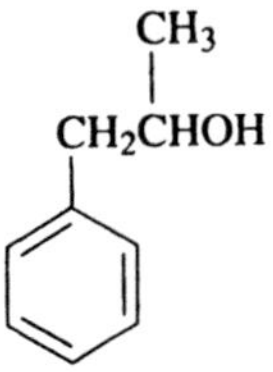

CH_3 CH_2CHOH

Carbinols (*continued*)

(tertiary alcohols)

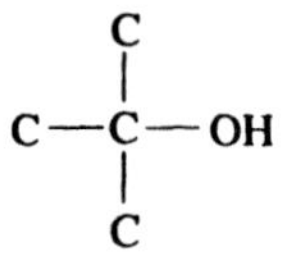

C—C(C)(C)—OH

Phenylethyl dimethyl carbinol

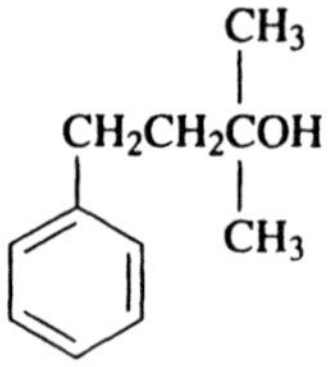

$C_6H_5CH_2CH_2C(CH_3)_2OH$

Cyclo hexyl

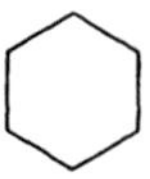

Allyl cyclohexyl propionate

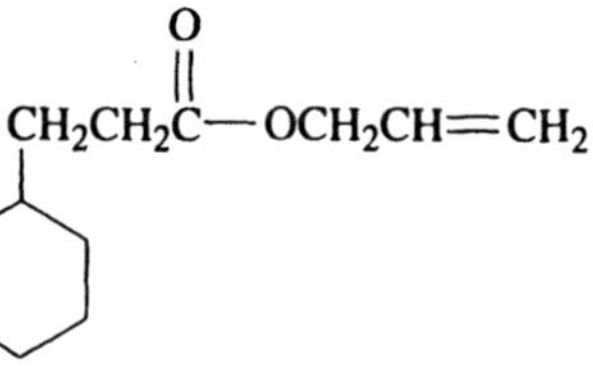

$C_6H_{11}CH_2CH_2C(=O)—OCH_2CH=CH_2$

Ionones (type of ketone) see Appendix D

Methoxy

$—OCH_3$

Ethoxy

$—OC_2H_5$

Phenols

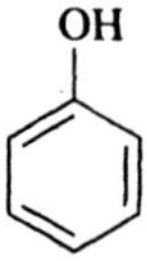

OH

Eugenol (4-allyl, 2-methoxy phenol)

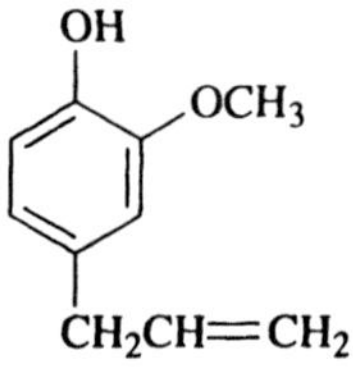

OH, OCH_3, $CH_2CH=CH_2$

Vanillin

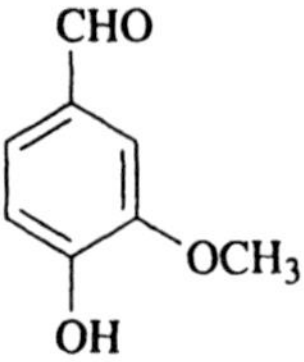

CHO, OCH_3, OH

Methyl salicylate

Pyran ring

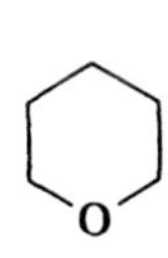

Rose oxide

Pyridine ring

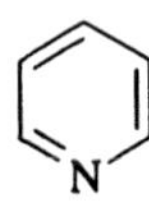

Iso butyl quinoline

Pyrrole ring

Indol

Appendix B

The Functional Groups

Family Name (Suffix)	Structure	Simple Example	Aromatic Example
Acid (-oic)	$-\overset{\displaystyle O}{\overset{\|}{C}}-OH$	$CH_3-\overset{\displaystyle O}{\overset{\|}{C}}-OH$ Acetic acid	$CH_2-\overset{\displaystyle O}{\overset{\|}{C}}-OH$ (on benzene ring) Phenylacetic acid
Alcohol (-ol)	$-\overset{\|}{\underset{\|}{C}}-OH$	$CH_3-\overset{\displaystyle H}{\overset{\|}{\underset{\underset{\displaystyle H}{\|}}{C}}}-OH$ or C_2H_5OH Ethyl alcohol	CH_2CH_2OH (on benzene ring) Phenylethyl alcohol

(see also Carbinols and Phenols under structural groups)

Family Name (Suffix)	Structure	Simple Example	Aromatic Example
Aldehyde (-al)	$-\overset{\displaystyle O}{\overset{\|}{C}}-OH$	$CH_3(CH_2)_xCHO$ Aliphatic aldehyde	CH_2CHO (on benzene ring) Phenylacetaldehyde
Acetal (-acetal)	$-CH\begin{matrix} \diagup O-\overset{\|}{C}- \\ \diagdown O-\underset{\|}{C}- \end{matrix}$ (the two C atoms joined by a vertical bond)		$CH_2CH\begin{matrix} \diagup OCH_3 \\ \diagdown OCH_3 \end{matrix}$ (on benzene ring) Phenylacetaldehyde dimethyl acetal

FAMILY NAME (SUFFIX)	STRUCTURE	SIMPLE EXAMPLE	AROMATIC EXAMPLE
Ester (-ate)	$-\overset{\vert}{\underset{\vert}{C}}-O-\overset{O}{\overset{\Vert}{C}}-\overset{\vert}{\underset{\vert}{C}}-$	$C_2H_5-O-\overset{O}{\overset{\Vert}{C}}-CH_3$ Ethyl acetate	$CH_2CH_2O\overset{O}{\overset{\Vert}{C}}CH_3$ (on benzene ring) Phenylethyl acetate
Ether (ether)	$-\overset{\vert}{\underset{\vert}{C}}-O-\overset{\vert}{\underset{\vert}{C}}-$	CH_3-O-CH_3 Dimethyl ether	$C_6H_5-O-C_6H_5$ Diphenyl oxide
Ketone (-one)	$-\overset{\vert}{\underset{\vert}{C}}-\overset{O}{\overset{\Vert}{C}}-\overset{\vert}{\underset{\vert}{C}}-$	$CH_3-\overset{O}{\overset{\Vert}{C}}-CH_3$ Acetone	$\overset{O}{\overset{\Vert}{C}}-CH_3$, CH_3, CH_3 (on benzene ring) Dimethyl acetophenone
Lactone (-one)	$-\overset{\vert}{C}-(CH_2)_x-\overset{\vert}{C}-$ $\vert \qquad\qquad \vert$ $O\text{———}C{=}O$	$CH_3(CH_2)_6CH-CH_2-CH_2$ $\vert \qquad\qquad \vert$ $O\text{———}C{=}O$ Undecalactone (Aldehyde C14— so called)	Coumarin
Nitrile (-nitrile)	$-\overset{\vert}{\underset{\vert}{C}}-C{\equiv}N$	$CH_3-C{\equiv}N$ Acetonitrile	CH_3, $C{\equiv}N$, $C{=}C$, CH_3, CH_3 (Terpenic) Geranylnitrile

FAMILY NAME (SUFFIX)	STRUCTURE	SIMPLE EXAMPLE	AROMATIC EXAMPLE
Nitro	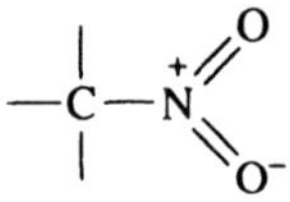 $-C-N^{+}(=O)-O^{-}$	CH_3NO_2 Nitromethane	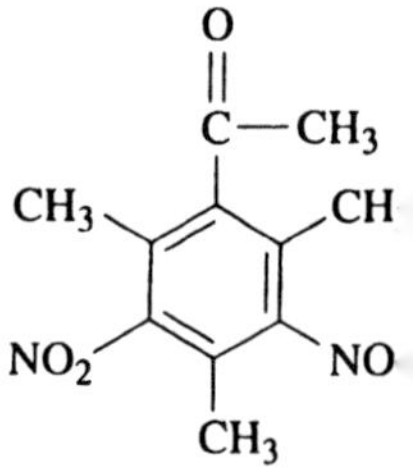 Musk ketone
Amine (-amine)	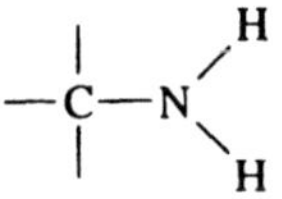 $-C-NH_2$	CH_3NH_2 Methylamine	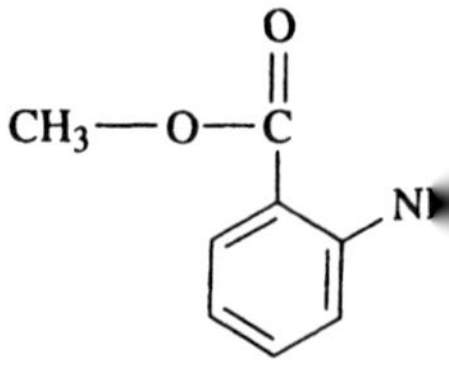 Methyl anthranilate
Sulphide (sulphide)	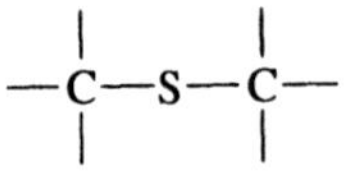 $-C-S-C-$	CH_3-S-CH_3 Dimethyl sulphide	

Appendix C

The Aldehydes

THE ALIPHATIC ALDEHYDES

Acetaldehyde

CH_3CHO

Saturated Straight Chain Aldehydes

$CH_3(CH_2)_xCHO$

Aldehyde C6	Hexaldehyde	
Aldehyde C7	Heptaldehyde	
Aldehyde C8	Octaldehyde	*Orange*
Aldehyde C9	Nonyl aldehyde	*Rose*
Aldehyde C10	Decyl aldehyde	*Citrus*
Aldehyde C11	Undecylic aldehyde	*Citrus-floral*
Aldehyde C12	Lauric aldehyde	*Soapy*
Aldehyde C13	Tridecyl aldehyde	*Waxy-citrus*
Aldehyde C14 (true)	Myrisitic aldehyde	*Fatty-citrus*

The so-called aldehydes C14 and C18 are not really aldehydes but lactones. Likewise, aldehyde C16 is really an ester.

Saturated Branched Chain Aldehydes

Aldehyde C11 — Methyl octyl acetaldehyde (MOA) — *Orange*

$$CH_3(CH_2)_7CH(CH_3){-}CHO$$

Aldehyde C12 — Methyl nonyl acetaldehyde (MNA) — *Herbal* (*Pine*)

$$CH_3(CH_2)_8CH(CH_3){-}CHO$$

Unsaturated Straight Chain Aldehydes

Aldehyde C11 — Undecylenic — *Rosy-fatty*

$$CH_2{=}CH(CH_2)_8CHO$$

Aldehyde C11 — Intreleven aldehyde — *Intense rosy*

$$CH_3CH_2CH{=}CH(CH_2)_6CHO$$

Leaf aldehyde — *cis*-3-Hexaldehyde — *Green*

$$CH_3CH_2CH{=}CHCH_2CHO$$

2,6 Nonadienal — *Violet-green*

$$CH_3CH_2CH{=}CH(CH_2)_2CH{=}CHCHO$$

THE TERPENIC ALDEHYDES

Monoterpenic Aldehydes

Citral
(two isomers)

Lemon

Neral

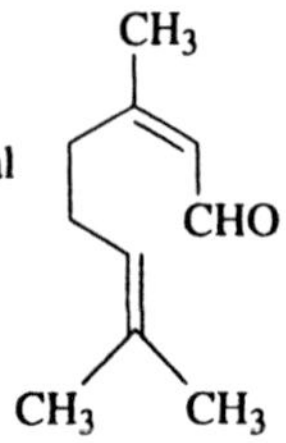

CH_3 | CHO | CH_3 CH_3

Geranial

CH_3 | CHO | CH_3 CH_3

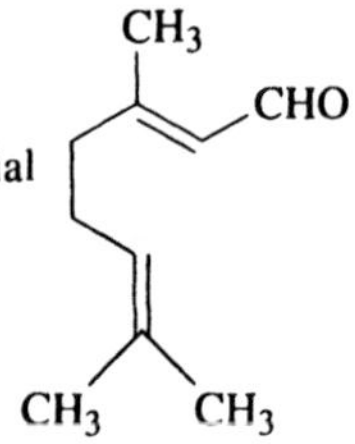

Citronellal

Citronella–citrus

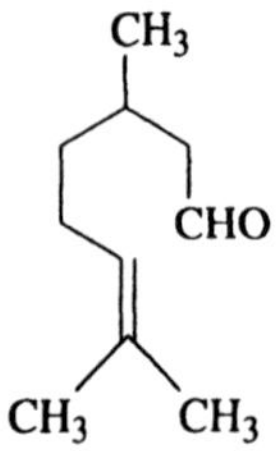

CH_3 | CHO | CH_3 CH_3

Hydroxycitronellal

Floral–muguet

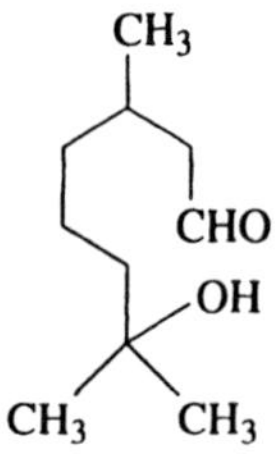

CH_3 | CHO | OH | CH_3 CH_3

Citronellyl oxyacetaldehyde
(muguet aldehyde)

Muguet–rose

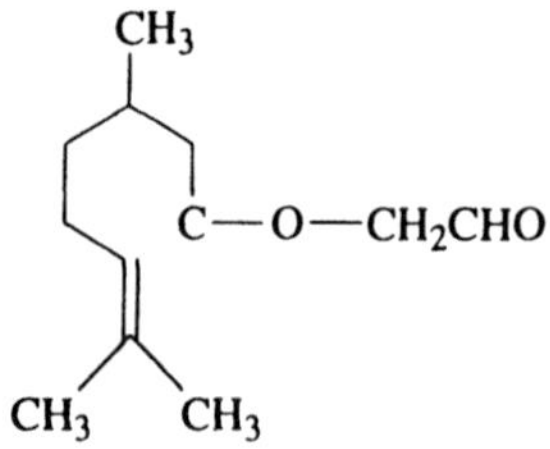

CH_3 | C—O—CH_2CHO | CH_3 CH_3

Norsesquiterpenic Aldehydes
(related to farnesol but with 14 carbon atoms)

Adoxal | Oncidal | *Floral*

THE NONAROMATIC CYCLIC ALDEHYDES

Triplal (two isomers)

Isocyclocitral (two isomers) *Green*

THE AROMATIC ALDEHYDES*

Nomenclature

The nomenclature of the aromatic aldehydes is somewhat confusing because in the past a number of different conventions were used in the naming of organic compounds. While some, such as cyclamen aldehyde, and vanillin, are usually referred to by their common names, others are referred to either by a single chemical name or by a number of different chemical synonyms.

This convention is best explained by taking as our starting point the three compounds, **benzene**, **phenol**, and **toluene**.

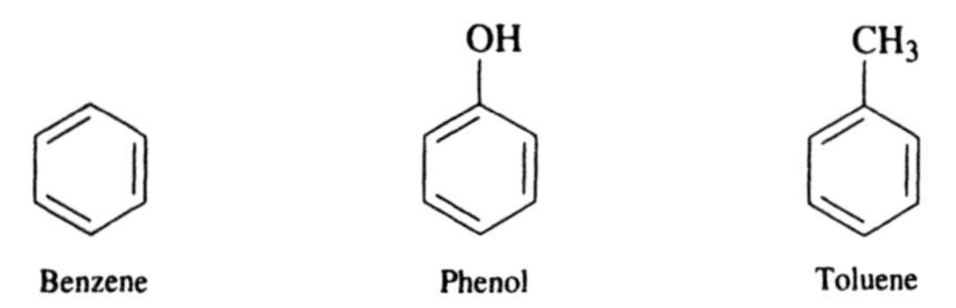

Benzene Phenol Toluene

*The term "aromatic aldehyde" is usually used in perfumery, as it is here, to include both the true aromatic aldehydes and the alkyl aromatic aldehydes, in which the aldehydic group is attached to a side chain rather than onto the benzene ring.

Relative to the hydroxy (OH) or methyl (CH_3) groups, the other positions on the ring may be named or numbered as follows:

OH, ortho, meta, para — OH, 2, 3, 4, 5, 6

Referring to the schematic representation of the aromatic aldehydes that follows, we may begin with **benzaldehyde** in which one hydrogen atom on the benzene ring is replaced by the aldehydic (—CHO) group. Moving down the page, we come to **phenylacetaldehyde** (phenylacetic aldehyde), so named as representing the condensation product of phenol and acetaldehyde in which one hydrogen atom of the methyl group is replaced by a benzene ring.

OH H—C(H)(H)—CHO → CH_2CHO + H_2O

Phenol Acetaldehyde Phenylacetaldehyde Water

Similarly **phenylpropyl aldehyde** may be considered as the product of phenol and propyl aldehyde. However, phenylpropyl aldehyde is also frequently referred to as either hydro-, or dihydro-, cinnamic aldehyde, owing to the fact that it was originally made by the hydrogenation of **cinnamic aldehyde**, an unsaturated aldehyde so named after its original isolation from cinnamon oil.

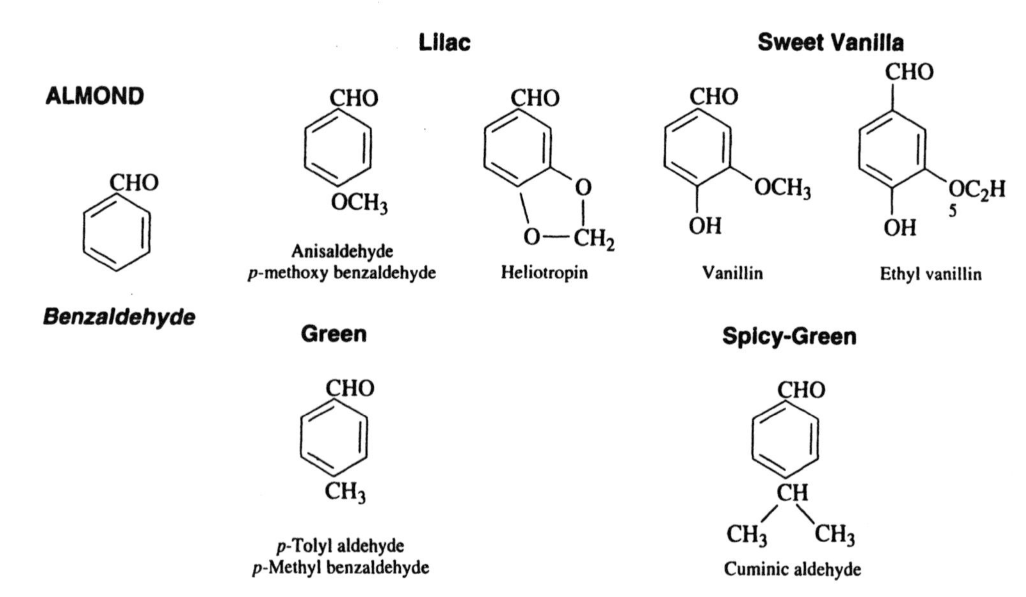

HYACINTH

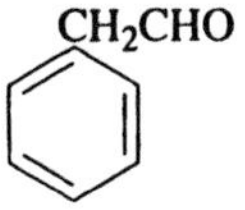

Green | **Leafy-Green**

CH_2CHO

Phenylacetaldehyde

CH_2CHO / CH_3

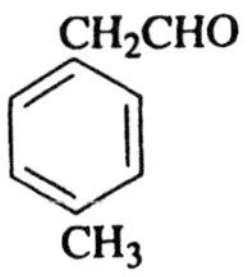

Syringa aldehyde
p-Methyl phenylacetaldehyde
p-Tolyl acetaldehyde

CH_2CHO / CH / CH_3 CH_3

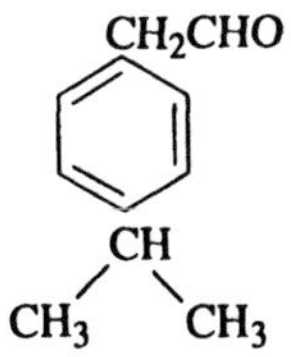

Cortexal
Homo-cuminic aldehyde
p-Iso propyl phenylacetaldehyde

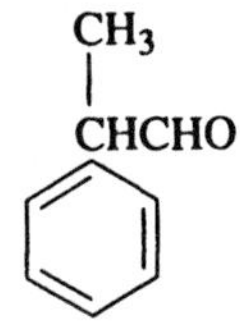

Hydratropic aldehyde

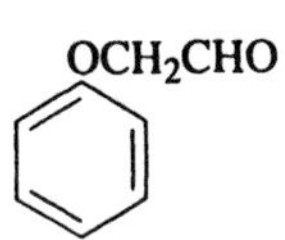

Cortexaldehyde

MUGUET

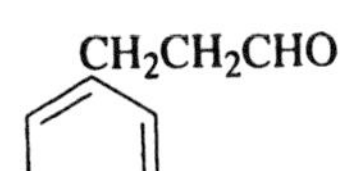

Phenylpropyl aldehyde
Hydrocinnamic aldehyde

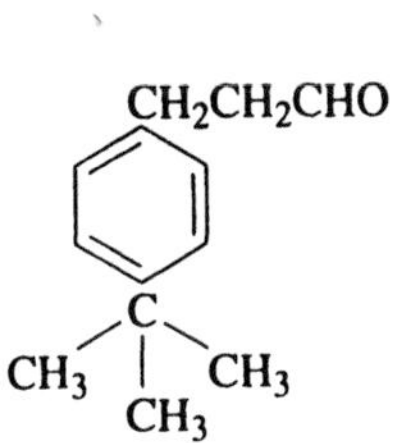

Bourgeonal

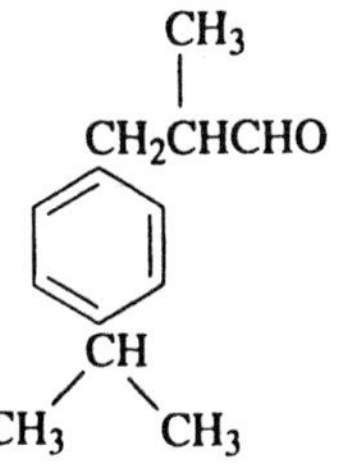

Cyclamen aldehyde

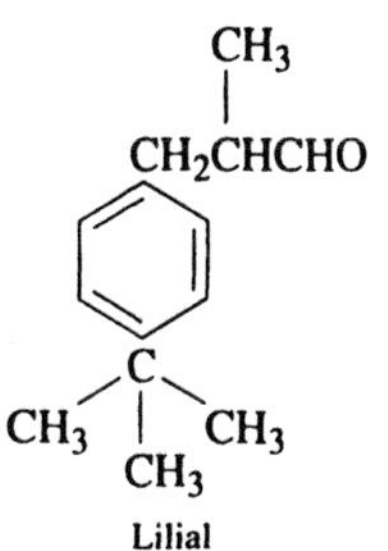

Lilial

CINNAMIC

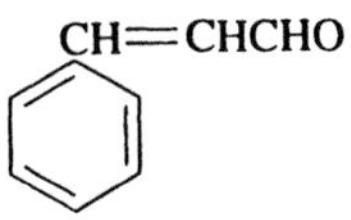

Floral—Jasmin

$CH{=}CHCHO$

Cinnamic aldehyde

C_5H_{11} / $CH{=}CCHO$

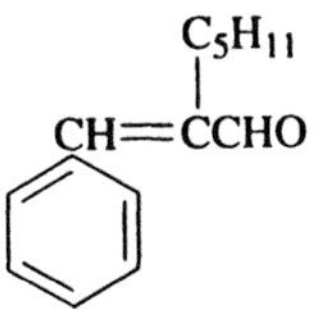

Amyl cinnamic aldehyde

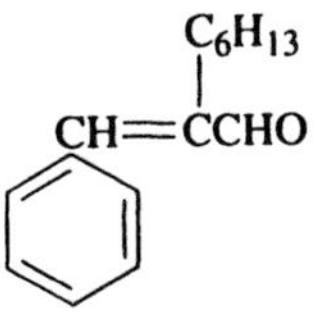

C_6H_{13} / $CH{=}CCHO$

Hexyl cinnamic aldehyde

Working across the preceding pages from left to right, we may derive from the four aldehydes at least one of the chemical names for each of the remaining products. (These are only shown when commonly used.) **Anisaldehyde**, for example, is sometimes referred to as *para*-methoxy benzaldehyde, and **vanillin** can be referred to as *para*-hydroxy-*meta*-methoxy benzaldehyde, or 4-hydroxy-3-methoxybenzaldehyde. (Not surprisingly the common name is the one preferred.)

Para-methyl benzaldehyde, however, is usually referred to as ***para*-tolyl aldehyde**, since it is equally correct to think of it as being the aldehyde derived from toluene. When referred to simply as tolyl aldehyde, this may denote that the material is a mixture of the *ortho, meta,* and *para* forms. Although similar in odor the *para* form is usually preferred. Similarly **syringa aldehyde** may sometimes be referred to either as *para*-methyl phenylacetaldehyde or as *para*-tolyl acetaldehyde.

Where a branched chain occurs in the molecule, this may also be reflected in the name. For example, **hydratropic aldehyde** may be referred to as alpha-methyl phenylacetic aldehyde—in this case with the methyl group occurring in the alpha position relative to the aldehydic group—and similarly **alpha amyl cinnamic aldehyde**. To give this material its even fuller name, we may refer to it also as alpha normal or alpha n. amyl cinnamic aldehyde denoting that the amyl ($—C_5H_{11}$) side chain is straight rather than in its branched, or iso form. Understandably perfumers prefer simply to call it ACA. To give two further examples, the chemical name for cyclamen aldehyde is alpha-methyl-*para*-iso-propyl hydrocinnamic aldehyde, whereas lilial is alpha-methyl-*para*-tertiary-butyl hydrocinnamic aldehyde.

To those unfamiliar with the complexities of chemical nomenclature, this lesson may all sound thoroughly confusing, but with a little practice it can be quite easily learned and applied to all types of aromatic compounds.

Appendix D

The Ionones

NOMENCLATURE

The nomenclature of both the ionones and methyl ionones can be confusing because they occur in a number of isomeric forms. Although it is possible to obtain many of these in a comparatively pure form, most of the products that are generally available contain a mixture of isomers besides the one that is named.

IONONE

Ionone Pure 100%

Products with this description are usually mixtures of ionones alpha and beta with the alpha form not less than 60%.

Ionones Alpha and Beta

These may be found in their pure forms or often containing a proportion of the other.

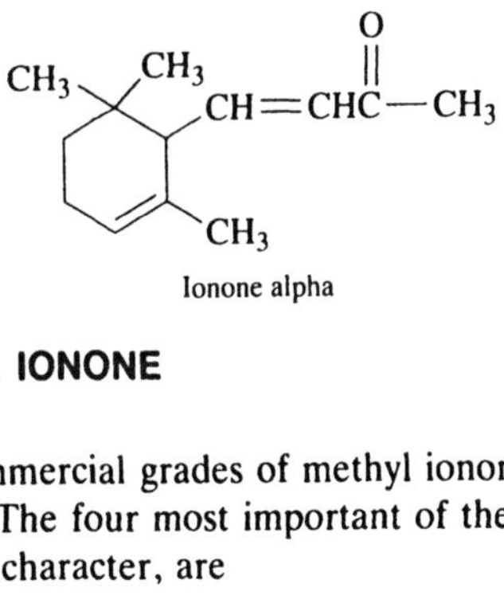

CH_3 CH_3 $CH{=}CHC({=}O){-}CH_3$ CH_3

Ionone alpha

CH_3 CH_3 $CH{=}CHC({=}O){-}CH_3$ CH_3

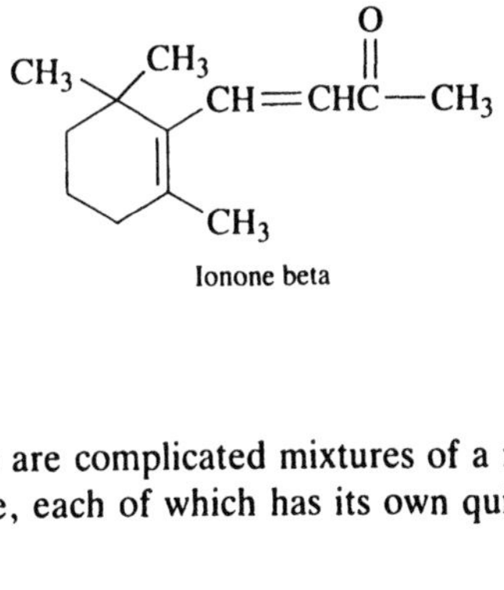

Ionone beta

METHYL IONONE

Most commercial grades of methyl ionone are complicated mixtures of a number of isomers. The four most important of these, each of which has its own quite distinct olfactory character, are

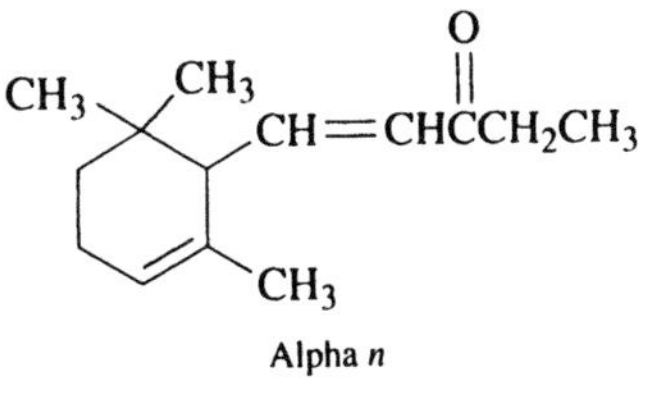

CH_3 CH_3 $CH{=}CHC({=}O)CH_2CH_3$ CH_3

Alpha *n*

CH_3 CH_3 $CH{=}C(CH_3){-}C({=}O){-}CH_3$ CH_3

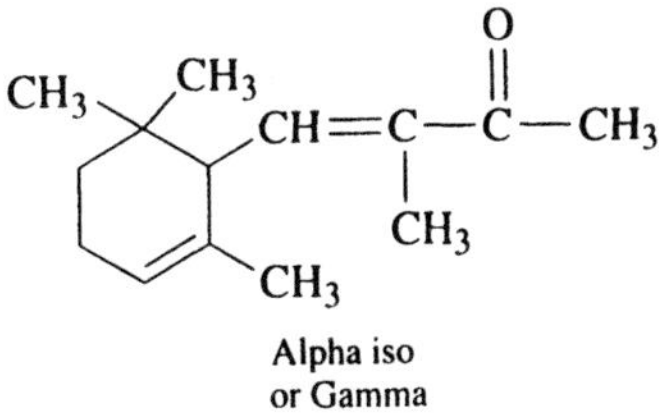

Alpha iso
or Gamma

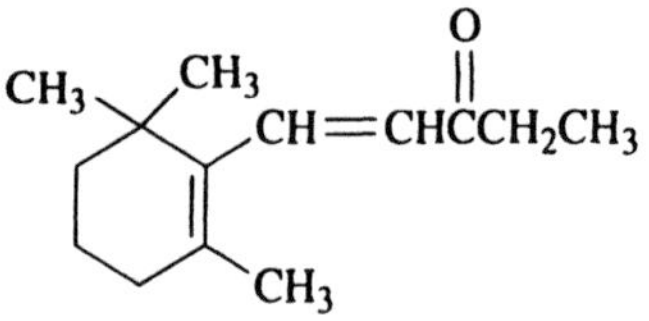

O
CH_3 CH_3 $CH{=}CHCCH_2CH_3$
CH_3

Beta *n*

O
CH_3 CH_3 $CH{=}C{-}C{-}CH_3$
CH_3
CH_3

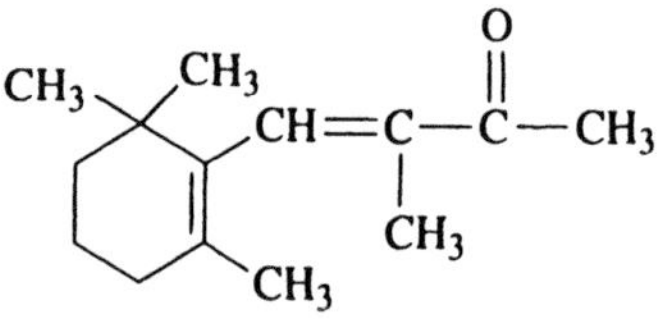

Beta iso
or Delta

For simplicity only the most commonly used names are given. Two other isomers often make up a large proportion of some of the less expensive grades of methyl ionone. These are

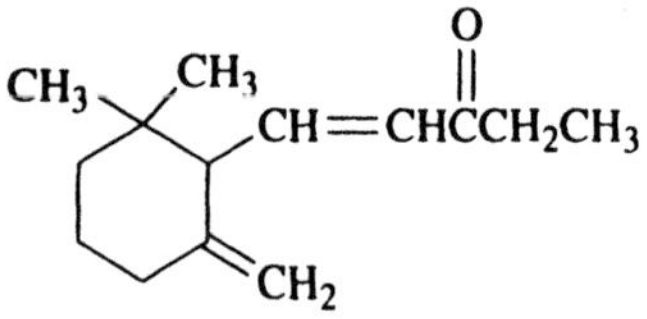

O
CH_3 CH_3 $CH{=}CHCCH_2CH_3$
CH_2

Gamma *n*

O
CH_3 CH_3 $CH{=}C{-}C{-}CH_3$
CH_3
CH_2

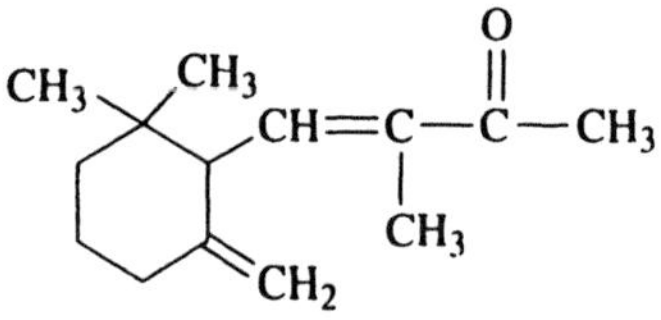

Gamma iso

Appendix E

Glossary of Perfumes, Speciality Raw Materials, and Bases*

Perfumes

Amarige	Givenchy	1991
Anais Anais	Cacharel	1979
Arpège	Lanvin	1927
Aramis	Lauder	1965
Aramis 900	Lauder	1973
Aromatics Elixir	Clinique	1972
Bandit	Piguet	1944
Beautiful	Lauder	1986
Blue Grass	Arden	1935
Cabochard	Gres	1958
Calandre	Rabanne	1968
Caleche	Hermes	1961
Calyx	Prescriptives Inc	1987
Canoe	Dana	1935
Casmir	Chopard	1991
Chanel No. 5	Chanel	1921
Chanel 19	Chanel	1971
Charlie	Revlon	1973
Chloé	Lagerfeld	1975
Chypre	Coty	1917
Coco	Chanel	1984
Coriandre	Couturier	1973
Diorella	Dior	1972
Dioressence	Dior	1970
Dune	Dior	1991
Eau Sauvage	Dior	1966
Escape	Calvin Klein	1991
Eternity	Calvin Klein	1988
Femme	Rochas	1942
Fidji	Laroche	1966
Fleurs de Rocaille	Caron	1933
Fougère Royal	Houbigant	1882
Giorgio	Giorgio	1981
Je Reviens	Worth	1932
Jolie Madame	Balmain	1953
Joop	Joop	1987
Knowing	Lauder	1988
L'Air du Temps	Ricci	1948
L'Heure Bleue	Guerlain	1912
L'Interdit	Givenchy	1957
L'Origan	Coty	1905
Loulou	Cacharel	1987

*The products in this list may not all still be available.

Perfumes (Continued)

Madame Rochas	Rochas	1960
Ma Griffe	Carven	1944
Miss Dior	Dior	1947
Mitsouko	Guerlain	1919
Moment Suprême	Patou	1933
Must de Cartier	Cartier	1981
New West	Lauder	1988
Obsession	Calvin Klein	1985
Opium	St. Laurent	1977
Oscar de la Renta	Stern	1976
Paloma Picasso	P. Picasso	1984
Paris	St. Laurent	1983
Poison	Dior	1985
Quelques Fleurs	Houbigant	1912
Rive Gauche	St. Laurent	1970
Samsara	Guerlain	1989
Shalimar	Guerlain	1925
Silences	Jacomo	1979
Spellbound	Lauder	1991
Topaze	Avon	1959
Trésor	Lancome	1990
Vanderbilt	Vanderbilt	1981
Vent Vert	Balmain	1945
Volupté	Oscar de la Renta	1992
White Linen	Lauder	1978
Youth Dew	Lauder	1952
Ysatis	Givenchy	1984

Specialty Chemicals and Derivatives

Agrunitrile	Dragoco
Ambroxan	Henkel
Brahmanol	Dragoco
Calone	Pfizer
Canthoxal	I.F.F
Cashmeran	I.F.F
Cassione	Firmenich
Cedramber	I.F.F
Celestolide	I.F.F
Cycloamylone	I.F.F
Damascone alpha	Firmenich
Damascone beta	Firmenich
Dimetol	Givaudan
Dragojasimia	Dragoco
Dupical	Quest
Dynamone	Givaudan Roure
Evernyl	Givaudan Roure
Frambinone	Dragoco
Fiorivert	Quest
Galaxolide	I.F.F
Glycolierral	Givaudan Roure
Hedione	Firmenich
Helional	I.F.F

Specialty Chemicals and Derivatives *(Continued)*

Hivertal (see Triplal)	Dragoco
Indolal (Florindal)	Dragoco
Isodamascone	Dragoco
Iso E super	I.F.F
Jessemal	I.F.F
Kephalis	Givaudan Roure
Lactoscatone	Dragoco
Liffarome	I.F.F
Lilial	Givaudan Roure
Lyral	I.F.F
Madranol	Dragoco
Magnolione	Givaudan Roure
Methyl cyclo citrone	I.F.F
Muguet aldehyde	I.F.F
Musk T (ethylene brassylate)	Tagasako
Oncidal	Dragoco
Parmavert	Bedoukian
Rholiate	Dragoco
Rosalva	I.F.F
Sandalore	Givaudan Roure
Sandranol	Dragoco
Sandela	Givaudan Roure
Tonalid	P.F.W
Triplal (see Hivertal)	I.F.F
Vertacetal	Dragoco
Vertenex (PTBCHA)	I.F.F
Vertofix	I.F.F
Vertral	Dragoco

Speciality Bases

Aldehone alpha	Roure
Althenol	Roure
Ambrarome	Synarome
Cassis 281	Dragoco
Dianthine	Firmenich
Dorinia	Firmenich
Jasmin 231	Firmenich
Florizia	Firmenich
Fleur d'Oranger	Firmenich
Melysflor	Firmenich
Mousse de Saxe	de Laire
Muguet Invar	Roure
Parmantheme	Firmenich
Pimenal	Roure
Printenyl	Roure
Selvone	Roure
Vertralis	Dragoco

Bibliography

Appell, L. 1964. Physical foundations in perfumery. *Amer. Perf. and Cosm.* **79** (1):25–33; **79**(2):43–48; **79**(5):29–41; **79**(11):25–39.

Arctander, S. 1969. *Perfume and Flavor Chemicals*. Publ. by the author, Montclair, NJ.

Becker, K., Koszinowsky, J., and Piringer, O. 1983. Permeation von Riech– und Aromastoffen durch Polyolefine. *Deutsche Lebensmittel-Rundschau* **79**:257–266.

Bell, H. 1985. The solubilization of perfumery materials by surface active agents. *Soap, Perf., Cosm.* **58**:263–273.

Burrell, J. W. K. 1974. The behavior of perfumery ingredients in products. *J. Soc. Cosm. Chem.* **25**:325–337.

Carles, J. 1961. Une méthode de création en parfumerie. *Recherches*, Dec. 1961. English transl. A method of creation in perfumery, *Soap, Perf., Cosm.* **35**:328–335 (1962).

Comité Français du Parfum and Société Française des Parfumeurs. 1990. *Classification des Parfums*. Paris.

Dervichian, D. 1961. Role de la structure moleculaire dans la fixation des parfums dans le savon. *La France et ses parfums*. Paris, p. 324.

Devos, M., Patte, F., Rouault, J., Laffort, P., and van Gemert, L. J. 1990. *Standardized Human Olfactory Thresholds*. IRL Press, Oxford.

Führer, H. 1970. The practice of composition. *Dragoco Report* **17**:3–13.

Gilbert, A. N., and Wysocki, C. J. 1987. The smell survey results. *National Geographic* **172**:514–525.

Haldiman, R. F., and Schuenemann, T. 1990. The hexagon of fragrance families. *Dragoco Report* **37**:83–89.

Hyman, A. M. 1977. Factors influencing the psychophysical function for odor intensity. *Sensory Processes* **1**:273–291.

Jellinek, J. S. 1959. The physico-chemical behavior of perfume materials in various carriers. *Am. Perfumer Aromat.* **73**(3):27–32.

Jellinek, J. S. 1961. Evaporation and the odor quality of perfumes. *J. Soc. Cosm. Chem.* **12**:168–179.

Jellinek, J. S. 1964. The effect of intermolecular forces on perceived odors. *Ann. N.Y. Acad. Sci.* **116**:725–734.

Jellinek, J. S. 1978. Fixation in perfumery—What we understand. *Perfumer and Flavorist* **3**(4):27–31.

Jellinek, J. S. 1979. The psychophysical function and the perfumer. *Dragoco Report* **26**:85–87.

Jellinek, J. S. 1990. A consumer oriented classification of perfumes. *Dragoco Report* **37**:16–29.

Jellinek, J. S. 1991a;. Perfumes and odors as a system of signs. In *Perfumes: Art, Science, and Technology,* P. M. Müller and D. Lamparsky, eds. Elsevier, London/New York, pp. 51–60.

Jellinek, J. S. 1991b. The impact of market research. In *Perfumes: Art, Science, and Technology,* P. M. Müller and D. Lamparsky, eds. Elsevier, London/New York, pp. 383–398.

Jellinek, J. S. 1992. Perfume classification: a new approach. In *Fragrance: The Biology and Psychology of Perfume,* S. Van Toller and G. H. Dodd, eds. Elsevier, London/New York, pp. 229–242.

Jellinek, J.S., and Warnecke, U. 1976. Faserhaftende Parfümierung von Wäscheweichspülern. *Seifen-Öle-Fette-Wachse* **102**:215–218.

Jellinek, P. 1954. *The Practice of Modern Perfumery*, transl. by A. J. Krajkeman. Leonard Hill Ltd., London.

Jellinek, P. 1993. *Die psychologischen Grundlagen der Parfumerie*, 4th edition, Hüthig Verlag, Heidelberg.

Lea, C. H., and Swoboda, P. A. T. 1958. Flavor thresholds of aliphatic aldehydes. *Chem. & Ind.*, London p. 1289.

Leffingwell, J. C., and Leffingwell, D. 1991. GRAS flavor chemicals—detection thresholds. *Perfumer and Flavorist* **16**(1):1–19.

Moreno, O., Bourdon, R., and Roudnitska, E. 1974. *L'intimité du parfum.* Perrin, Paris.

Muller, P. M., Neuner-Jehle, N., and Etzweiler, F. 1993. What makes a fragrance substantive? *Perfumer and Flavorist* **18**(4):45–49.

Naves, Y. R. 1957. Natural odorants and their synthetic reproduction: Past, present, and future. *Soap, Perf., Cosm.* **30**:1140–1144.

OECD. n.d. *OECD Guidelines for Testing of Chemicals,* Organization for Economic Cooperation and Development. Paris.

Patte, F., Etcheto, M., and Laffort, P. 1975. Selected and standardized values of suprathreshold intensities for 110 substances. *Chem. Senses and Flavor* **1**:283–305.

Pickthall, J. 1956. An approach to soap perfumery. *Soap, Perf., Cosm.* **29**:808–813.

Pickthall, J., 1974. Perfumes and colour reactions in soaps. *Soap, Perf., Cosm.* **47**:311–320, 342–356.

Poucher, W. A. 1955. A classification of odours and its uses. *J. Soc. Cosm. Chemists* **6**(2):80; *Amer. Perf. and Ess. Oil Review*, July 1955:17–24.

Roehl, E. L., and Knollmann, R. 1970. Olfactory and analytical control of the odour-effect of perfumes in various cosmetic preparations. Proceedings of the Joint Symposium on Perfumery of the BSP and SCC of Great Britain, Eastbourne, Sussex. Also *Naarden News* **22** 1971:4–7.

Roudnitska, E. 1962. The young perfumer and scents. *Dragoco Report* **9**:83–113.

Roudnitska, E. 1991. The art of perfumery. In *Perfumes: Art, Science, and Technology,* P. M. Müller and D. Lamparsky, eds. Elsevier, London/New York.

Saunders, H. C. 1973. An approach to fitting a perfume to the polarity of its substrate. *Cosmetics and Perfumery* **88**(11):31–34.

Sauvageot, F. 1987. Differential threshold and exponent of the power function in the chemical senses. *Chem. Senses* **12**:537–541.

Sfiras, J., and Demeilliers, A. 1957. Molecular structure and organoleptic quality. S.C.I. Monograph **1**:9.

Sfiras, J., and Demeilliers, A. 1964. Étude par chromatographie gazeuse de la vapeur odorante émise par une savonette parfumée. *Recherches* **14**:33–43.

Sfiras, J., and Demeilliers, A. 1966. Rendement parfumant d'une savonette déterminé par chromatographie. *Recherches* **15**:87–88.

Sfiras, J., and Demeilliers, A. 1974. Étude par chromatographie gazeuse de la vapeur odorante émise par une savonette parfumée. *Recherches* **19**:259–268.

Shiftan, E., and Feinsilver, M. 1964. Practical research of the art of perfumery. *Ann. N.Y. Acad. Sci.* **116**:692–704.

SRI International. 1992. *Flavors and Fragrances*. Menlo Park, CA, p. 117.

Stevens, S. S. 1957. On the psychophysical law. *Psychol. Rev.* **64**:153–181.

Stoddart, M. 1990. *The Scented Ape*. Cambridge University Press, Cambridge.

Stravinsky, I. 1942. *Poetics of Music*, transl. by A. Knodel and I. Dahl. Harvard University Press, Cambridge, MA.

Streschnak, B. 1991. Support materials for odorant mixtures. In *Perfumes: Art, Science, and Technology,* P. M. Müller and D. Lamparsky, eds. Elsevier, London/New York, pp. 347–362.

Sturm, W., and Mansfeld, G. 1975. Haftung und Fixierung von Riechstoffen. *Chemiker-Ztg*. **99**(2):69–78; Engl. transl. Tenacity and fixing of aromatic chemicals, *Perfumer and Flavorist* **1**(2):6–15 (1976).

Wells, F. V. 1960. Le Comportement des parfums sur la peau. *Parf. Cosm. Savons* **3**(1):4–10.

Index of Perfumery Raw Materials and Bases

At pages marked in *italic* type, the chemical formula is given.

General Index

Perfume names are in *italic* type.

Printed in the United States
141175LV00002BA/9/A

9 780471 589341